THE ENTIRELY ACCURATE ENCYCLOPAEDIA OF EVOLUTION

ROBERT NEWMAN

FREIGHT
BOOKS

First published in the UK 2015

Freight Books
49–53 Virginia Street
Glasgow, G1 1TS
www.freightbooks.co.uk

Co-published with Cargo Publishing

A CIP catalogue reference for this book is available from the British Library.

ISBN 978-1-908885-53-1
eISBN 978-1-910449-62-2

Typeset by Freight in Plantin

Printed and bound by Hussar, Poland

the publisher acknowledges investment from
Creative Scotland toward the publication of this book

for Simon Lewis & Sophie Allain

LIST OF CONTENTS

INTRODUCTION

The Entirely Accurate Encyclopaedia of Evolution sets out to trace the forces which, over millions of years, have worked to put monkey and human at different ends of the barrel organ.

This book grew out of a stand-up show called *A New Theory of Evolution*. Two passions have governed the writing of both show and book. Delight in the living world's mind-boggling multiplicity, and horror at the hijacking of evolutionary theory by ideologues with a very weak grasp of Darwinism. And I'm not the only one upset by this sort of thing.

In *The Selfish Gene*, Richard Dawkins castigates an 'erroneous explanation,' which he laments is 'even taught in schools.' It's the calamitous idea that: 'living creatures evolve to do things [...] "for the good of the species" or "for the good of the group."'

Authors who believe this, he says, 'have got it [natural selection] totally and utterly wrong.'

Here's an extract from the sort of author he means:

> *In social animals [natural selection] will adapt the structure of each individual for the benefit of the community.*

The author is Charles Darwin, and the book is *On the Origin of Species*, which I'm afraid is even taught in schools.

In 1871, when his book on human evolution *The Descent of*

Man came out, Darwin was still getting his own theory 'totally and utterly wrong':

> *We can perceive that an instinctive impulse, if it be in any way more beneficial to a species than some other or opposed instinct, would be rendered the more potent of the two through natural selection.*

It gets worse. The very least you'd expect from a scientist is for him to reject the use of subjective terms like *love, sympathy* and *feeling.* Abstract nouns like these are unquantifiable, unscientific and, frankly, girly. You would hope a famous naturalist would focus instead on measurable material data. You would hope in vain. Deplorably, Darwin uses the word *love* 133 times in *Descent of Man.* Not until Lennon and McCartney do you find another Englishman willing to bang on about love quite as much as Darwin does. If *love* wasn't bad enough, it grieves me to tell you that in the same book Darwin uses the word *sympathy* 64 times, and *feeling* an atrocious 79.

Nor is this an aberration. In *The Expression of the Emotions In Man and Animals,* Darwin clocks up a Personal Best 84 uses of *feeling,* then speculates that language may have evolved from singing love songs.

Reading Darwin is the very best medicine against the neo-Darwinists, who for the last 40-odd years have told us that a more or less ruthless duplicity lies behind all human behaviour. This has given us a disastrously narrowed and pessimistic view of what it means to be human – for proof of which look no further than any drama-documentary purporting to reconstruct life among the early humans. Any *Walking With Hominids* series always features the same shaggy-pelt-wearing troglodytes making these weird gnashing, growling vocalisations expressive of resentment, frustration, mutual suspicion and rage. It's the sound you'd get if you shook all the consonants out of the *Daily Mail.*

Happily, recent science is returning Darwinism back to Darwin. The dog-eat-dog version of evolution has had its day, but the damage it has done will take some fixing. I hope that *The Entirely Accurate Encyclopaedia of Evolution* may be a small part of the repair kit.

A

ANT

'The brain of an ant is one of the most marvellous atoms of matter in the world,' wrote Darwin, 'perhaps more so than the brain of man.'

And yet ants are only intelligent in a colony. 'Nobody has ever been able to teach an individual ant anything.'[1] A lone ant will make a random decision at a Y junction every single time even if the eats and treats are *always* in the same place. If the brain of an ant is an atom, then it needs the field-effect of other brains for its atomic charge.

AUSTRALOPITHECUS LUCY

A million years ago in the Late Pleistocene, the population of early humans was dwindling. We'd dwindled down to a couple of tribes, or possibly down to one single solitary pregnant individual Australopithecus Lucy. This is the Bottleneck Theory, which rests on the fact that everyone alive on earth today shares DNA with Lucy. What we know is that she left Africa, and then in some remote corner of Europe, Lucy's sons and daughters, I'm afraid to say, copulated with each other and produced offspring. Their offspring also copulated with each other. But then their offspring moved out of Norfolk, spread out and scattered over continents, flourishing spectacularly as they became more diverse.

1 Brian Goodwin, *How The Leopard Changed Its Spots: The Evolution of Complexity*, Phoenix (1997).

Indeed in 2013, the oldest human footprints ever found outside of Africa were discovered in Happisburgh, Norfolk, but there is a mystery here. The 800,000 year old Happisburgh footprints have baffled paleoanthropologists because they are too ordered to have been produced by beachcombing or samphire picking, yet not quite linear enough to be walking in a straight line either.

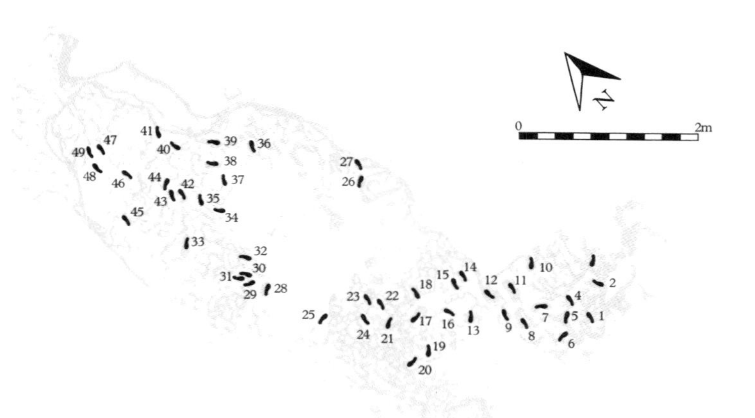

Me and a team of paleoanthropologists from Stanford University mapped the footprints using photogrammetry, ran the pattern through a supercomputer database and made an astounding discovery. The Happisburgh footsteps exactly match only one other known pattern of human footsteps:

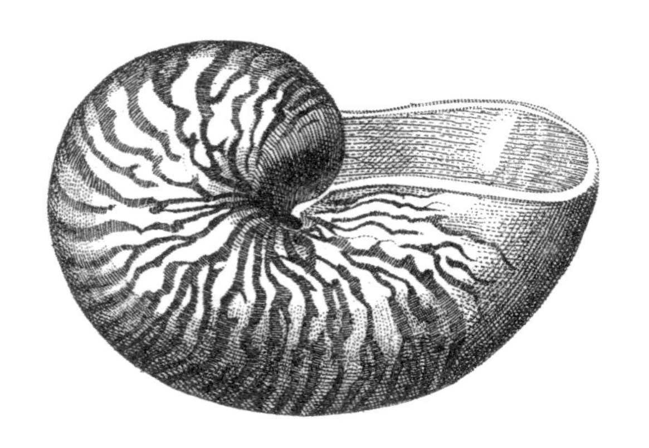

B

BABOON

As dawn broke on a Maasai Mara tourist lodge rubbish tip, a battle raged between two groups of olive baboons. Forest Troop was fighting Talek Troop for a plastic bag full of raw beef burgers. Forest Troop's war party won the battle, but the meat was tainted with bovine tuberculosis and the troop's entire warrior class died of TB. At a stroke, Forest Troop lost all its most aggressive, high-ranking males. The ratio of females to males in the troop doubled overnight.

In their paper *A Pacific Culture among Wild Baboons: Its Emergence and Transmission* ethologists Robert Sapolsky and Lisa Share describe the cultural shift that followed.[2] The baboons began to maintain troop cohesion by mutual hip-hugging and grooming rather than biting and fighting. Not only were things more peaceful within the tribe, no more war parties were sent out to attack other tribes either.

When new adult males migrate in from neighbouring selfish groups, Forest Troop baboons induct these incomers into the post-TB ethos. This induction process explains how this ethos has survived for a quarter of a century, generation after generation, despite all kinds of new males arriving in from other troops.

Sapolsky and Share's findings have helped put the kybosh on that cherished 70s dogma about the universal principle of selfishness. *Selfish Gene* readers were told that even given the 'improbable chance existence' of such a group of altruists, nothing would 'stop selfish individuals migrating

2 Robert M Sapolsky and Lisa J Share, *A Pacific Culture among Wild Baboons: Its Emergence and Transmission*, PLoS Biology Vol 106 (2004).

in from neighbouring selfish groups, and by inter-marriage, contaminating the purity of the group.'

But that's not what happens in real life, it turns out. In fact, as Sapolsky and Share show, what happens among real live animals is the complete opposite.

Hormone samples show Forest Troop's baboons to be less stressed than those in the hierarchical, violent Talek Troop. The olive baboons are happier now things are more equal. As distinguished ethologist Frans de Waal said:

'If baboons can do it then why not us? The bad news is that you will first have to knock out all the most aggressive males.'

Is there a non-violent way to do this? I think there is, and its name is Virgin Galactic.

In our time the aggressive males who are bending society out of shape are not so much the warrior class as plutocrats, oligarchs, and corporate bosses. Here's my plan of how to free ourselves from their malign control.

When Virgin Galactic finally gets off the ground, here's what we do. We wait until Richard Branson and all his multi-millionaire and billionaire chums are orbiting in space, and just before they emerge from the dark side of the moon, we get everyone around the whole world to switch off all the lights. And now they are lost in space!

We knock out their satnav, and remotely activate the long transparent pole we secretly attached to their fuselage before blast-off. The long transparent pole extends forwards. At the end of it is an illuminated classroom globe. When the billionaires spot it, they go:

'Look! There it is! The blue planet! Set a course for home!' The more they head towards this globe the further into deep space they travel. With every 10,000 miles the globe comes another inch towards the windscreen – just to give them a sense of achievement – until at last it touches the nose-cone of the rocket and pops open to reveal a screen showing footage of Branson's mansions being used as social housing for key-sector workers.

C

CYCLOSA TREMULA

Cyclosa tremula is a black and white striped Guyanan spider who does something wonderful. She builds dummy spiders with which she populates her orb web. She makes these eight-legged replicas from the husks of dead insects she has caught. Because they are made from prey debris they're a dull grey colour. The local birds soon learn not to eat these disgusting, tasteless grey spiders. When a black and white striped Cyclosa sees a bird coming, she bounces up and down on her web, blurring her stripes, and when you blend black and white you get grey, so the birds fly away.

Now, it is important to remember here that we have *absolutely no idea* why Cyclosa makes these replica spiders. She *may* be making them as cunning decoys, or she may not. We do not know. And the simple fact of our near-perfect ignorance is too easily forgotten.

In *Darwinian Populations and Natural Selection*, philosopher of science Peter Godfrey-Smith points out that because human minds are powerful when it comes to thinking in terms of strategies and tactics, ruses and ploys, we too quickly attribute this type of motive to other organisms. The tendency to think other organisms are always plotting and scheming he calls 'Darwinian paranoia'. We think the spider has designs and cunning stratagems when – perhaps – she has none.

But why else could she be making these dummy spiders if not as a clever decoy? Well, I was reading one book which had a marvellous throwaway phrase, it just said:

> *Of course it may be that Cyclosa builds her replica spiders out of loneliness.* [3]

Spider loneliness! What a concept! I love this concept because it means that loneliness confers a selective advantage. How does nature select for loneliness? It's simple. A Cyclosa who never feels the need to build dummy spiders for company is more likely to get picked off by an orange-bellied sparrow. It's just like my mum used to say to me when I was a kid:

'Unless you feel a very strong sense of loneliness the birds will eat you.'

But if our black and white Cyclosa is building the spiders out of loneliness, you say, then what about the bouncing up and down? Doesn't that prove that it's all a decoy strategy just like in the Second World War when Churchill ordered the British Army to make tens of thousands of cardboard tanks which were then placed in fields in Kent?

All these dummy tanks were painted in green and brown camouflage markings, their guns were made from the cardboard tubes from inside old wallpaper rolls. As well as cardboard tanks there were papier-mâché Spitfires, and dummy barracks made of crepe paper stretched between giant willow hurdles. The reason for doing this was because Churchill knew that Luftwaffe spy planes were taking aerial reconnaissance photographs. The hope was that when these photos were developed in Berlin, the German High Command would take one look at them and say:

'There's no point trying to invade Britain, the whole country's made out of craft materials. Those huge piles of coal we hoped would fuel our war machine turn out to be pebbles painted black and lightly dusted with a glitter shaker.'

So isn't Cyclosa building her decoy spiders in the same way as the British built dummy tanks and planes? Doesn't the fact

3 F John Odling-Smee, Kevin N Laland and Marcus W Feldman, *Niche Construction: The Neglected Process in Evolution,* Princeton University Press (2013).

that she bounces up and down when birds are near prove that their whole purpose is to confuse birds?

Could be. Or it could be that like the Guyanan bird-eating spider *Theraphosa blondi*, she fancies her chances, and when she sees an orange-bellied sparrow swooping in for the kill, she is on her toes giving it that pre-fight bounce:

'Come on, I'll have the whole flock of you!'

Maybe she fancies her chances so much that she is bouncing up and down trying to get lift-off. She wants to bounce into mid-air, meet the bird eye to eye and say:

'Come round here again and I'm gonna build my next spider out of sparrow debris.' Or she could just bounce up in front of her wearing a feathered headdress.

My personal hunch? I reckon she is bouncing up and down to feel the reassuring pressure of the web pushing back against the pads of her feet. It's reassuring because it's always worked before. She doesn't know why, but for some reason whenever there's trouble overhead, if she just bounces up and down for a bit the trouble soon flies away. And so what nature has selected for her is loneliness coupled with neurotically bouncing when scared.

This reminds us of a crucial fact: natural selection and progress are two very different things. The way selection works is like this. Whatever's left at the end of the week, for whatever jammy, ass-backwards reason, is the stuff from which variation will come. If you have survived by being more lonely and more deluded about how scary you look when you bounce up and down, then that is every bit as exquisite an evolutionary strategy as having sharp claws or opposable thumbs.

Evolution is not onwards and upwards. Just onwards. As Darwin once wrote: 'there is no up or down'. Unless, of course, you're bouncing among cadaverous effigies on a homemade trampoline!

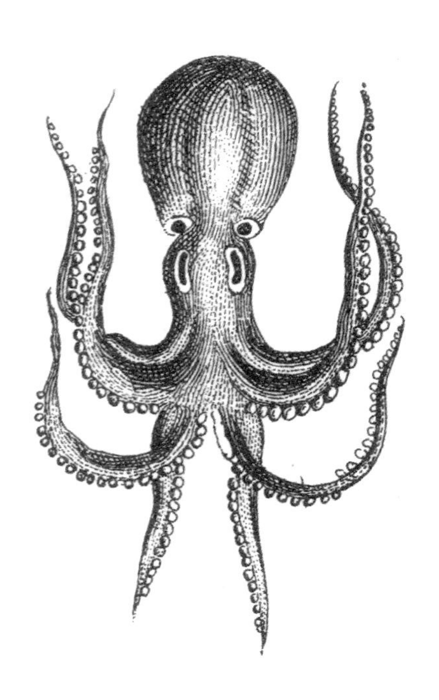

D

DICTYOSTELIUM DISCOIDEUM

Current dogma holds that selection happens only for selfish reasons of reproductive fitness increase, never for any generous impulse that could in any way lower a gene's frequency. This view has, alas, swept all before it in our time: every public intellectual, every science journalist and almost every university faculty in the land. There are, however, some impressive holdouts. One group that have never bought into this doctrine are the amoebae Dictyostelium Discoideum.[4]

Unicellular forest floor foragers, they live in colonies of between 10 – 100,000 subsisting on a diet of the bacteria they find in leaf-litter mulch, rotting logs, and decaying bark. But in time of famine, the Dicty Discos perform an astonishing sequence of transformation. They first emit a chemical signal – cyclic adenosine monophosphate – which brings them in from all parts of the forest to one central location. Once gathered together, they decide to spend the afternoon turning themselves into a single multicellular slug with light-sensitive eyespots and a tail. As this single slug they writhe, squirm and wriggle their way towards a patch of sunlight. When they get there, the first 20% form a rigid stalk out of their own bodies, up which the other 80% climb to form a giant fruiting berry at the top. And just as the breeze disperses dandelion puffball spores, so a gust of wind bursts the fruiting berry and the Dicty Discos go floating through the air like Allied parachutists over Occupied France – and with a roughly equal chance of being rescued by a member

4 I am indebted for this take on Dictyostelium Discoideum to Oren Harman's wonderful book, *The Price of Altruism: George Price and the Search for the Origins of Kindness,* Vintage (2011).

of the French Resistance in that nation of enthusiastic Nazi collaborators. One of the great myths of twentieth century history is the existence of a French Resistance. Here's a statistic you don't hear as often as you should: there were actually less members of the French resistance than of Dexy's Midnight Runners.

Scandalously inefficient, extravagantly far-fetched, wildly impractical though this evolutionary strategy be – it works! More often than not the Dicty Discos land where there is more of the delicious, nutritious bacteria they find in leaf-litter mulch and decomposing timber, begging the question WHERE in their unicellular, amoeboid brains do they retain the memory that this is something that they can all do together in time of crisis, in time of need?

Another question is what about the 20% that form the rigid stalk? What happens to them? They do not get to fly off to bacterial pastures new. They die that others may live. Individuals sacrifice themselves for the group.

DIRECTED PANSPERMIA [AKA CRICK THEORY]

In 1973, Sir Francis Crick, as in Watson and Crick, proposed a whole new theory of evolution called Directed Panspermia, which he set out in these words:

> *'Microorganisms travelled in the head of an unmanned spaceship sent to earth by an advanced civilisation which developed elsewhere billions of years ago. Life on earth started when these organisms were dropped into the primitive ocean and began to multiply. [...] It may emerge eventually that it would have been impossible to start life on earth [in any other way], whereas on some other more favourable planet, life could have started more easily and evolved more rapidly.'*[5]

5 Francis Crick, *Life Itself*, Simon & Schuster (1973).

He then writes, *brilliantly:*

> *'Our unusual moon may turn out to have been more of a handicap than an advantage.'*

... And he gives not a word of explanation! He just says it and he's off, distracted by the chimes of a passing ice-cream van, or the opening credits of Star Trek, leaving what must surely be one of the most extraordinary sentences in the entire English language just hanging there in the sky like a glowing silver orb.

> *'Our unusual moon may turn out to have been more of a handicap than an advantage.'*

Well, the moon has been many things to many people but, I ask you, an *impediment*? An *obstacle*? You have a couple of builders come round to do a kitchen extension.

'Oh, this'll be a very straightforward job,' says the gaffer. 'You got a nice flat level surface here. Ease of access for materials, oh yeah, this'll be a lovely job, we'll have this done in no – *(looks up at sky)* Oh fuckin' 'ell! Who put that there? No, mate. No, no, no, mate. The gravitational pull's gonna ripple the plaster. Put the tools backs in the van, Lance.'

Francis Crick's 1973 attempt to deny our complete and utter dependence on the earth was echoed the following year when Richard Dawkins described our immortal genes as 'sealed off from the outside world' (as if there were no such thing as heritable hazing of genes by exposure to radiation or to harmful chemicals such as the fungicide vinclozolin).[6] Now, these two totally un-Darwinian, grandiose, narcissistic fantasies have surely contributed to the extreme difficulties now being faced by atmospheric scientists in trying to get people to attend to

6 Randy L Jirtle and Michael K Skinner, *Environmental Epigenomics and Disease Susceptibility,* Nature (2007).

what it means that we have disrupted the Polar Jet Stream.[7]

Ordinarily, the Polar Jet Stream is an egalitarian superhero. She flies around the world equalizing its weather systems. Too much sun? She'll push some rain clouds your way. Too much rain? She'll push those rain clouds out to sea. But lately it's all been getting a bit too much for her, and when it does she just goes on a six week bender in a pub in Anglesey, drunkenly slurring the words:

'I'm not going anywhere for anybody. I'm staying right here. And I'm going to make a very big puddle. It's all got a little bit too much. The EU has just reclassified gas as a green fuel. Now to my mind... to my mind... that is the most sinister reclassification since Neil Diamond was inducted into the Rock 'n' Roll Hall of Fame. He's a Jazz Singer. By his own admission. It's a Hydrocarbon... It's all got a little bit too much. I'm not going anywhere for anybody. Let the good people of the Somerset Levels float down the High Street on rafts made from their Top Gear DVD's! Goodbye-ee!'

Crick Theory banishes the earth as the site of evolution, nominating instead 'some more favourable planet,' where 'life could have [...] evolved more rapidly.'

But *why* would life want to evolve rapidly? What's its hurry? Natural selection has no sense of future. It is not hoping for anything. For two billion years blue-green algae were on a loop tape.

What I believe Crick means when he talks about life evolving more rapidly is that his humanoid aliens were trying to arrive at us more quickly, because we are the spit of them, right down to our seed banks and rocket ships. Underlying the sci-fi, then, is a Biblical vision of creation:

> *'Furthermore God said, let us make men in our image, according to our likenes[s].'*[8]

7 In the psychological sense of narcissism as a projection of omnipotent fantasies upon an outside world which has ceased to exist except as a mirage, or extension of the ego.

8 *Genesis 26*, Geneva Bible (1560).

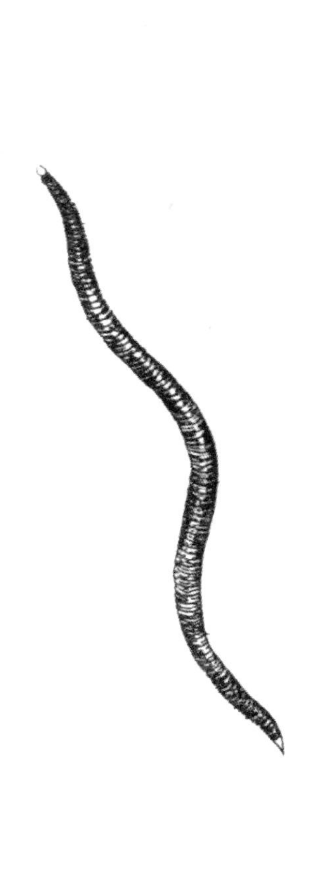

E

EARTHWORMS

Humans are an earthworm-dependent species. But for earthworms, no single acre of wheat, barley or corn would ever have been harvested.

The sad and pitiable people who dream of space colonies cannot grasp the simple fact that there is no soil in space. Worms cannot live on regolith, and if they're not going, we're not going.

EPIGENETICS

At the start of the twenty-first century, at McGill University, Montreal, Michael Meaney and Moshe Szyf led a team of biologists who discovered that when Mamma Rat licks and nuzzles her pups' heads she changes the gene expression of their **hippocampus.**

Turns out that when mamma rat licks and nuzzles her pups' heads and necks she stimulates the production of proteins called transcription factors, which slide down the bass and treble on the gene responsible for release of the stress hormone cortisol, so that when startled these rats are much less stressed. (I believe it was just such a laidback rat that I caught in the act of scoffing my packed lunch on the banks of the River Avon in Wiltshire earlier this year, because when the rat looked up and saw me, he didn't bat an eyelid:

'I've started without you but do join me,' his look seemed to say. 'I am Rat Boy, natural custodian of nuts and cheese. I've had a paddle in the margarine but not so you'd notice. I see you're fond of sun-dried tomato and foccacia. From Hebden Bridge are you?')

What they did next at McGill was switch litters. They took high-stressed, high-cortisol rat pups, and had them fostered by a nurturing, licking and nuzzling Mamma Rat. The results were mind-blowing. Not only did these foster pups have the changed gene expression and reduced cortisol release, but – and this is the really amazing thing – so did their offspring and their offspring's offspring! It turns out, after all, that certain intense life experiences can be passed down in the chromosome from generation to generation. A terrifying thought!

Just last year, out of Emory University, Atlanta, Georgia, came an astounding study of epigenetic effects in mice, likely to be true of all mammals, including ourselves. Turns out, if you take a smell that Grandpa Mouse learns to associate with fear, the first whiff of the same smell will terrify his offspring's offspring and their offspring too! This was very much in my mind Friday before last when I was in the back of a late-night minicab that was being driven extremely recklessly. Swinging from the rear-view mirror was a very pungent pine air-freshener. But I tried to control my fear because I don't want my grandchildren to feel edgy in a coniferous forest.

Epigenetics is non-DNA heredity. The discovery that proteins *outside the gene* transmit heritable traits to daughter cells has debunked the notion that we are the passive playthings of devious molecules, and put paid to the hoary old dogma about how the organism is just a gene's way of making a copy of itself, a vehicle for 'tyrannous replicators' to replicate themselves. If epigenetic changes can be passed down without so much as a by-your-leave from the gene, if the all-powerful tyrant turns out to be the last to know what the organism is up to, then Selfish Gene Theory's melodramatic language of 'tyrant replicators' no longer fits anything we find in the living world. Although it remains an excellent synopsis of most interactive PC/video role-play fantasy games: immortals use humans as proxies in an eternal war.

EVOLUTIONARY PSYCHOLOGY

Evolutionary Psychology is the name of the craze for explaining human behaviour by what we were doing in the Stone Age.

A classic example of Evolutionary Psychology is Hurlbert and Ling's paper *Biological Components of Sex Differences In Colour Preference*, which claims that the reason girls like pink and boys like blue has evolutionary roots in the Pleistocene epoch, when boys were hunters and girls were gatherers.[9]

For a boy, blue skies mean good hunting. Blue is the colour of the water hole where he will spear his first megatherium. For a girl, pink is the colour of ripening berries, the cranberry, the strawberry, the raspberry – try not to think about the blueberry or the whole argument falls apart.[10] And because of this division of labour, evolution has *hardwired* girls to like pink and *hardwired* boys to like blue.

But neuroscientist and philosopher Raymond Tallis points out that until the late Victorian era it was the other way around. Blue for girls (think Alice In Wonderland's dress) because of its associations with the Virgin Mary, and pink for boys because it's 'a pale version of ferocious red.'[11] Not until 1869 do we find a character in Louisa May Alcott's novel *Little Women* deciding to try out,

9 AC Hurlbert and Y Ling, *Biological Components of Sex Differences In Colour Preference,* Current Biology (2007).

10 Blank from your mind also the blackberry, bilberry, sloe berry, blackcurrant, loganberry, dewberry, huckleberry or plum.

11 Raymond Tallis, *Aping Mankind: Neuromania, Darwinitis and the Misrepresentation of Humanity,* Routledge (2014).

> *'the new French fashion of putting blue ribbons on boy babies and pink ones on girl babies.'*[12]

1869! Not exactly the 2 million years-ago Pleistocene Epoch, is it? But maybe this means we evolved from Victorians! Hard-wired into our neural circuits, therefore, is a love of railways, iron bridges and prostitutes with a heart of gold.

12 From the Pitt Rivers Museum's childhood collection notes. These also argue that the nineteenth century pink and blue fashion was facilitated by the mass production of synthetic aniline fabric dyes.

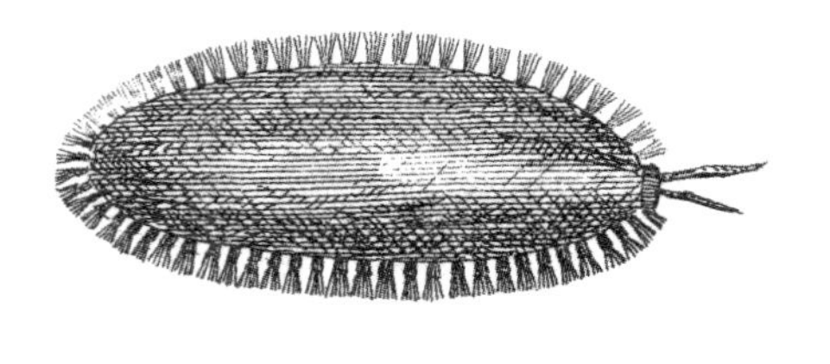

F

❧ FOSSIL RECORD ❧

The lack of intermediate species in the fossil record is a headache for Darwinism. The fossil record doesn't show gradually evolving species, as the theory of natural selection predicts it should. Instead the same fossils are repeated layer upon layer, over and over again, as free of innovation as the subterranean Terracotta Army of the First Chinese Emperor. After aeons of undeviating stability, a new species will suddenly appear, fully formed. As a recent editorial in *Transactions of the Paleontological Society* put it:

'If you're looking for phyletic gradualism, there's fuck all.'

The fossil record is like CCTV footage missing the key moment. Here's the suspect coming out of the train station, there he goes entering the park, then you lose him for a while, and next time you see him he's hovering on iridescent wings while shooting out a sticky tongue three times his own body length.

Gaps in the fossil record leave the evolution of flight a particular mystery. The sudden appearance in the fossil record of the first feathered, flying bird, Archaeopteryx, is like the way Joy Division burst onto the early 80s music scene: both seem to have come from nowhere, there's nothing else like them, and nothing will ever be the same.

If you go way back in the fossil record, there are lizards using their lateral scales for thermo-regulation, scales that may or may not have evolved into wings. Then the fossil record shows nothing, nothing, nothing, until – BOOM! – Archaeopteryx.[13] This jump cut raises some problems. 5% of a wing won't fly. So

13 Or Aurornis xui, a new candidate for first bird.

let's says Archaeopteryx evolved from a lizard that used to use its flappy bits of hanging down arm skin to cool itself – when does bingo wing become actual wing? 'Structural continuity, functional discontinuity,' is Stephen Jay Gould's answer. Or in other words, you have to have everything in place but be using it for something else like when you use the old kitchen sink for a birdbath, same thing, different use. That's why *The Entirely Accurate Encyclopaedia of Evolution* believes flight evolved accidentally from the habit of 'puffing up'. Frigate cobras, frilled lizards, toads, robins, great horned owls and elephant seals all puff themselves up. (Right from the off, I think we can safely rule the elephant seal out of our enquiries into the origin of flight.) For birds such as robins or great horned owls, puffing up is a way of keeping warm by trapping air under the feathers, and also a way of looking bigger.

If flight came from puffing yourself up so as not to be bullied, then that first lift-off must have been the best pre-fight psyche-out ever. Imagine it. One minute you're a flightless proto horned-owl, nervous and afraid. A huge flightless alpha-male proto horned-owl is in your face shouting,

'Wanna make something of it, do ya? Huh? Huh?'

Slowly you ascend into the sky. The bully is so psyched out that you can afford plenty of backspin on the words as you casually enquire:

'You want some?'

G

GENDER ROLES IN CHILDCARE

The dogma that gender roles in childcare are genetically determined came to the fore in 1975, simultaneous with the passing into law of the Sex Discrimination Act. And by the way, whenever any of those instant history, *How We Made Modern Britain*-type programmes mention the 1975 Sex Discrimination Act, it's always presented as though it had just somehow slipped the gentlemen's minds to accord women their employment rights, but no sooner were the gentleman made aware of the ladies' desire to work than they fell over themselves to correct this unintended slight.

'Oh, I had no idea! So you want to work, do you? Splendid, splendid. How many of you? All of you? I'm not sure you've thought that through. I'm not sure there's enough florists in the entire country!'

And lo! It did come to pass that the menfolk of these islands did raise unto the heavens a mighty wail of lamentation:

"GENDER EQUALITY HAS GONE TOO FAR IN THE OTHER DIRECTION! GENDER EQUALITY HAS GONE TOO FAR IN THE OTHER DIRECTION!"

Cometh the hour, cometh the male primatologist striding from the jungles of Congo and the tropical forests of Borneo clutching evidence – gleaned from his study of baboon – to prove that gender roles in childcare are genetically determined.

'We have observed,' Desmond Morris or someone equally appalling would tell the BBC TV cameras, 'that the male baboon can never be entirely sure which of the offspring he himself has fathered. Now, the reason for his uncertainty is because a very high proportion of adult female baboons are scrubbers. Therefore the male baboon's best chance of increasing his inclusive fitness, of ensuring that his genes are the ones passed onto successive generations, is to become the dominant alpha male, with a harem of submissive females, a Mick Hucknall tour-bus scenario.'

At the end of the 1970s the first female primatologists went into the jungles of Congo and the tropical forests of Borneo, where they discovered baboon troops not organised around dominant alpha males but around female kin networks, where new males can only join the troop if they know a female member. And even then, they must serve a probationary period during which they prove their worth... how? By hunting? By fighting? By killing? No, by foster care! *Specifically looking after offspring not their own, who do not share their DNA.*

Now, readers I know you're thinking:

'Newman, you've got all this from some anarcho-feminist pamphlet printed with menstrual ink, and you're about to tell us that far from being a display of aggression, when the male baboon beats his chest he's empathising with his partner's mastitis.'

No. Everything from the foster-care to the female kin networks comes from David Attenborough *Trials of Life*, Disc 1, Episode 2, *Growing Up*. Now it has cost me something to tell you this because I don't want you thinking that I've just cobbled this book together from DVD box sets bought in charity shops. And so to prove to you that I have, in fact, been doing rigorous academic research in esoteric journals and not just cribbing all this off the telly, I should now like to cite a study that was published in a recent edition of a journal called *Transactions of the Paleontological Society*. It's a fascinating study, which actually has a bearing on this very issue, because what the paleontologists did was they excavated a Stone Age quarry – they quarried a quarry, if you like – and using fossil DNA were able to build up quite a detailed picture of the day to day workings of this Neolithic quarry. And it turns out that Stone Age Man may have been using a long-necked brontosaurus as a type of crane. And when the site-foreman tugged a pterodactyl's tail to signal the end of the shift, the crane-operator would slide down the brontosaurus's neck, fly off the tail, hit the ground running and sprint all the way back to the cave, where he would be served a juicy T-Rex steak.

Now of course, what William Hanna and Joseph Barbera were nostalgic for wasn't the Neolithic era, but the 1950s before Betty and Wilma went to work.

I should declare an interest in all this, which has to do with a scientific study published last year that got national news coverage. You may have heard it yourself. Scientists were claiming to have discovered that men who do an unusually large amount of childcare have unusually tiny testes. Now, I have two small problems with this theory.

In 1950s Britain the number of husbands doing equal childcare as their wives was very nearly zero. Does this mean, therefore, that the loss of Empire and heavy industry has been accompanied by the decline of bollocks? Or does it just feel that way?

❧ GENE ❧

In *The Selfish Gene,* Richard Dawkins writes that:

> *'gene selfishness will usually give rise to selfishness in individual behaviour.'* [14]

But no organism's behaviour mimics how its strands of DNA go about their business. Fidgety polymers do not beget fidgety elephant seals. The dolphin's polypeptide chains jerk in uneven spasms even as she describes a graceful arc. The pangolin's simple life does not reflect the complex cellular machinery of her protein molecules. 'The inference from [gene] function to overt psychology,' as Simon Blackburn puts it, 'is simply not available.' Crime scene DNA does not give you motive. Only a man suffering from extreme mental illness would look down a microscope at his wife's squiggly nucleotides and mutter:

'So, that's how it is! She is out to get all she can from me and then – *vamoosh!* Just as I suspected! Ho-ho! Well, two can play at *that* little game.'

Question the idea that genes control destiny, however, and you will be told that you can't handle hard, cold reality, and are seeking more or less theological shelter. A cosy and wholly illusory universe awaits those unable to stare into the void with the fearless intensity of intellectual heroes such as Steven Pinker, EO Wilson, and Richard Dawkins.

Might I suggest that, faced with the terrifying randomness of life, an idea of genetic destiny offers its own consolation? In the face of the accidental death and arbitrary fates that await us what could be more comforting than a belief that you are born with your destiny inscribed inside you on a 2-metre long scroll of nucleic acid wrapped tight around the histone? What could

14 From time to time Dawkins claims to disavow this doctrine. And yet with every new edition of *The Selfish Gene* he reprints it again. Why not strike out what you admit to be false? Why privilege the well-received popular myth over what the science says?

be more reassuring than the neo-Darwinists' *Bourne Identity* view of human nature? Plonk us down anywhere in the world, and we will still come out the same, because it is written in our genes. It's in our DNA.

Now, before I criticise the theoretical underpinnings of the Matt Damon trilogy – *The Bourne Identity, The Bourne Supremacy* and the *Bourne Ultimatum* – I should declare an interest. Readers of *Screen International* will already know about the severe public humiliation I suffered in connection with the Bourne franchise. For those readers who don't know, let me bring you up to speed. Studio bosses at Universal pulled the plug on production of my original screenplay *The Bourne Ontogeny,* which was to have been the second in the trilogy. After just a fortnight of shooting, Universal halted production, sacked the crew and went with rival script *The Bourne Supremacy.* Here, in outline, are the only seven scenes we shot.

1. NIGHT. EXT. Jason Bourne floating face down in sea, rescued by Italian fishermen.

2. NIGHT. INT. Inside the boat, Bourne joins half-drowned migrants from Syria, Somalia, Eritrea, and Libya.

3. DAY. INT. Lampedusa, Immigrant Detention Pen. Finding he can speak English, Jason Bourne acts as interpreter for fellow refugees, and lands a job with a refugee rights NGO.

4. DAY. EXT. Downtown Lampedusa. Sitting at a cafe piano, Bourne finds to his delight that he plays piano to Grade 8. Troublingly, however, he seems only to know the songs of Carole King. 'Was I thrown off a cruise ship?' he wonders. Discovering himself also to be an unusually resourceful and tenacious man, Bourne scours junk shops and picks up a second-hand djembe, accordion, battered acoustic guitar, Casio keyboard, zukra (Libyan bagpipe) and qanum (Syrian

harp). After auditioning fellow refugees Bourne forms the Immigrant Orkestra of Lampedusa.

5. NIGHT. INT. Eurovision Song Contest. Bourne's composition *Immigration Built The Nation* wins for Italy!

6. DAY. INT. A courtroom in Rome. Lawyers acting for the Carole King estate allege that Bourne's *Immigration Built The Nation* is lifted note for note from *(You Make Me Feel Like) A Natural Woman.*

NIGHT. EXT. Jason Bourne found floating face down in sea.

Summoned to Universal HQ, studio bosses told me they would never have greenlit production in the first place had they known how far *Ontogeny* departed from *Identity.*

'Aha,' I replied, 'but that's where you're wrong: ontogeny *is* identity! Ontogeny is your life history, and as I told Matt, but for accidents of history, you see, those immigrants could be you or me. No intrinsic properties of DNA determine who is found helplessly floating on the water and who is not. That's the whole message of the film.'

'We're going with *Supremacy.*'

To have no truck with the credo that DNA is destiny (which Israeli professor Eva Jablonka brilliantly calls 'genetic astrology'!) is not to deny genes a role in natural selection. In fact, among the strongest critics of selfish gene theory are those whose work first reconciled genetics and Darwinism, such as Ernst Mayr.

In the 1930s and 40s the Modern Synthesis rescued natural selection from semi-obscurity by demonstrating that Mendelian genetics powerfully corroborated Darwin and Wallace's theory of evolution. (Up to then, it had been thought that Mendelian genetics actually *disproved* Darwinism.) The leading lights of the Modern Synthesis were JBS Haldane, Theodosius Dobzhansky, August Weissmann, Ronald "Piggy" Fisher, Sergei Chetverikov, Sewall Wright and Ernst Mayr. In 1999, Ernst Mayr gave an interview in which he said:[15]

> *Dawkins' basic theory of the gene being the object of selection is totally un-Darwinian [...] and totally impractical. A gene is never visible to natural selection, and in the genotype it is always in the context of other genes, and it's the interaction with those other genes that makes a particular gene either more favourable or less favourable. In fact, Dobzhansky worked on so-called lethal chromosomes which are highly successful in one combination but lethal in another. Therefore people like Dawkins in England who still think the gene is the target of selection are evidently wrong.*

If genetic determinism really were true, argued the great American evolutionary biologist Stephen Jay Gould, then the fact that it's ugly would be neither here nor there. 'It would be just another of those distressing biological facts of life, like death and disease, to which we must, alas, accommodate ourselves.' But the fact that it's a fallacy, he says, makes it tragic that this has become the dominant mythology of our time, the fairytale we tell each other about who we are and our place in the world.

The danger of putting the wrong storyteller at the head of the table was something Charles Dickens learnt the hard way when he invited Hans Christian Andersen to come and stay indefinitely, having never met the man in his life. This is a true story, and apparently things began to go badly wrong at the very first supper of his stay with the Dickenses in Doughty Street, Bloomsbury. There was Dickens, his wife Catherine, their ten kids. Head of the table sits Hans Christian Andersen, who, at the end of the main course, says:

15 Edge.org. 31.12.1999.

HANS: Pass me that sausage please.

DICKENS: I believe you've just had six sausages.

HANS: Yes, and now I would like that sausage, please, which will make seven.

DICKENS: Ah, bit of a problem there, you see, I was actually hoping to have that sausage for my lunch tomorrow. But perhaps I can interest you in some plum pudding and custard for desert?

HANS: So, you refuse me the sausage. Boys and girls, would you like to hear the story of the greedy old King who passed a law forbidding any of his subjects to eat sausages?

DICKENS: Have the sausage.

HANS: No, no. I tell the story. One morning the King was sitting down to his breakfast. Can you guess what it was? That's right! A great big steaming cauldron full of piping hot sausages! Oh, said the King, dancing around with the cauldron in his arms, 'all the sausages are mine! Nobody else can have any except for me coz I'm the King. Court Chamberlain, do you want a sausage? You can't! It's against the law! The sausages belong to me! I am the sausage King!'

DICKENS: Have the fucking sausage!!!

After Hans Christian Andersen had been staying with the Dickenses for six long months, host and guest were about to go for a walk across Coram's Fields, when Dickens had to turn back at the last minute to comfort and console one of his sons Sydney, who was traumatized by a story that Hans Christian Andersen had told him. It was the story of *The Wicked Little Boy*

Whose Eyes Were Pecked Out By Ravens Because He Wouldn't Make Hot Chocolate For The Danish Visitor.

The moral of the story is that for the last 40 years the neo-Darwinists have been the silver-tongued houseguests at the head of the table, holding us spellbound with simplistic melodramas which, upon inspection, turn out to be mere projections of peculiarly selfish natures upon the world.

GENETICALLY MODIFIED CROPS

Some of the same people who champion GM foods also say they believe in evolution. Me, I don't see how you can believe in both. One or the other, yes, but not both.

In 2010, Monsanto admitted to the journal *Science* that pink bollworm, a rampant insect pest, had evolved resistance to biotech's flagship crop Bt cotton. This is cotton modified to carry a gene from Bacillus thuringiensis (Bt) which was supposed to kill insects stone dead, but which pink bollworm appears to quite like, or at least not much mind.[16] (Pink bollworm is a **nematode** and there's a whole entry in this encyclopaedia on how fiendishly clever they are).

Since then, a study in *Nature Biotechnology* has found that of 13 major pest species, five have evolved a resistance to Bt corn or cotton. It's taken them a decade.

Once a pest has figured out how to pick a crop's genetic lock, they can wipe out whatever you plant. This is a particular weakness for GM crops because they are monocrops, planted in vast single stands, which makes them more – not less – vulnerable to pests. As such, genetically modified crops are, I would argue, a major threat to food security.

A review of 15 years of biotech agriculture published

16 Pallava Bagla, *Hardy Crop-Munching Pests Are Latest Blow To GM Crops,* Science Vol 327 (2010).

in the journal *Progress in Physical Geography* declared that: 'current GM approaches are relatively transitory as a means of combating pests.' Meanwhile biotech is winging it with desperate stopgap measures such a releasing swarms of sterile insects into the air in the hope that they will mate with resistant crop pests and check their proliferation. Not all farmers can afford to do this. Even those who can must ruefully reflect, when they're standing on a tractor trailer in a field of manky cabbage next to a crate of aphids eunuchs, how far they've come from those hi-tech graphics of a modified gene being spliced into RNA sequence as colourful as a row of beads.

The arguments for GM foods are like the arguments for high-rise flats and inner-city motorways in the 1950s. Merrily bulldozing half-timbered Jacobean alms-houses to make way for town-centre flyovers, 1950s architects and town councilors were dismissive of the outdated human-scale make-do-and-mend approach. These were all to be ploughed under and replaced with the 'modern'. Decades later, we are still spending billions to repair the damage that sci-fi modernism did to town and city-centres. But it won't be nearly so quick and cheap to put the superbugs, superpests, and superweeds back in the bottle, once we've poisoned the once-rich diversity of birds and insects.

Bent on sowing the Southern Hemisphere with GM seeds, Bill Gates is the King Canute of the paddi-field, thinking he can stop natural selection with a few well-aimed transposons and patented petro-chemical inputs. But nothing stops natural selection.

From China comes a cautionary tale of what happens when you kill off all the pests in favour of farming monocultures. Sparrows eat rice grains, and so in 1958, Chairman Mao launched his Kill A Sparrow campaign as part of the Great Leap Forwards. Millions of citizens set about catching and exterminating sparrows. Within a year the only sparrows left in China were disguised as parakeets and living under assumed names on the Kazakhi border. Killing all the sparrows, however, led to a plague of locusts and a catastrophic collapse in rice harvests. The Kill A Sparrow campaign was abandoned in 1960.

Still, no-one would ever let a lone megalomaniac impose his parascientific vision on millions of hectares again, right?

Non-GM successes, meanwhile, never get any coverage, so let's hear it for: allergen-free peanuts, salt-resistant wheat, heat and drought resistant beans, and virus-resistant cassavas. Not long after farmers start swapping and sharing these seeds, all these new hybrids become common property.

But where's the profit in that?

❧ GROUP SELECTION ❧

In 1776 Benjamin Franklin told the Second Continental Congress in Philadelphia:

'We must all hang together or assuredly we shall all hang separately.'

Expressed in terms of evolutionary biology, Franklin's insight goes like this:

'Social cooperation among its members increases a group's fitness above the arithmetic mean of the individual members' fitness.'[17]

Charles Darwin held it to be self-evident that group selection is a strong force in evolution, but nowadays the Church Scientific condemns as heresy the idea that nature selects for mixed-ability groups rather than just successful individuals. According to current dog-eat-dog dogma, heavily influenced as it is by Thatcherite No-Such-Thing-As-Society doctrine, group selection shouldn't happen in theory and so doesn't happen in practice. But that hasn't, of course, stopped it going on just as vigorously as ever.

17 Ernst Mayr, *What Evolution Is,* Basic Books (2002).

Take a colony of red harvester ants for example. The individual ants don't make more ants. The colony makes another colony. That's because, strictly speaking, it's not a population, it's a collection of sterile sisters.[18] Yes, I grant you, they do keep a couple of males in a wall-cupboard and they bring them out every few years when they want to make another colony. You can recognise the male red harvester ant by his tiny head – he's only going to live for two weeks which isn't long enough to eat so he's got no jaw musculature. The other reason for his tiny head is he's only got a tiny brain, because all he has to do at the end of that two weeks is mate with the queen. As he withdraws, his entrails are ripped out of his body, and, with his last gasp, you can just hear him say:

'Gender equality has gone too far in the other direction!'

18 Deborah Gordon, *The emergent genius of ant colonies,* TED talk (2003).

H

❧ HIPPOCAMPUS ❧

The hippocampus is the bit of the brain that's used for (among other things) navigation, path finding, orientation, and sense of direction. A famous study of London cab drivers found that the hippocampus is larger in them than it is in the rest of us. The same study also found that there's another part of London cabbies' brains – the amygdala – that's slightly smaller than it is in the rest of the population. I submitted a paper to the science journal *Nature* arguing that this part of the brain is where a positive attitude towards multiculturalism is located.

❧ THE HOSPITALS ARE COMING ❧

Archaeologists and anthropologists appear at last to have reached a consensus about what Stonehenge was for. It's now thought that Stonehenge was a hospital. Turns out the reason for dragging all that bluestone from Preseli Hills to Salisbury Plain was because of the healing properties in its dolerite and rhyolite. Paleopathology corroborates this hypothesis. A 2009 dig near the henge excavated the bones of the Amesbury archer, who is thought to have traveled all the way from Switzerland for treatment on a badly damaged knee. What has really clinched the argument that Stonehenge was a prehistoric NHS is radiocarbon dating, which has been able to pinpoint not just that the Amesbury archer arrived at Stonehenge in 2300 BC, but also that he wasn't actually seen by a doctor until 2275 BC.

I wish more people knew the story of Stonehenge General.

If they did we'd have a bluestone rampart against the idea that human evolution stops once you have hospitals. 'We stopped natural selection,' David Attenborough told the *Radio Times* in 2013, 'as soon as we started being able to rear 90-95% of our babies that are born. We are the only species to have put a halt on natural selection of our own free will.'

Now with great respect, some trepidation and no credentials to back me up save for a single name check in the thanks and acknowledgements section of one solitary paper published this year in science journal *Nature*, I am now going to disagree with one of the world's great naturalists, David Attenborough.[19]

Consider the earthworm. If an earthworm is born into a soil with a pH made less acidic by its worm ancestors, such that weak or sickly newborns that would never have survived the hard scrabble soil of yesteryear now live, does that spell the end of earthworm evolution? Is life too soft for earthworms now that the pH balance of the soil gives them a better start in life? Are the old Earthworms telling the Young 'Un:

OLD EARTHWORM: You've had it too easy. You don't know what it was like. When I was born there was a car battery lying in that flowerbed. There were small boys here in this back garden too pulling me in half. You couldn't hack it, gel. You'd be six feet under.

YOUNG 'UN: We are six feet under, Dad.

OLD EARTHWORM: You'd be pushing up the daisies.

YOUNG 'UN: We are pushing up the daisies, Dad.

OLD EARTHWORM: I mean, you'd be in a grave.

YOUNG 'UN: Living the dream.

19 S L Lewis & Mark A Maslin, *Defining the Anthropocene,* Nature 519 (2015).

If you find yourself unfit for your local habitat, then instead of letting natural selection pick you off, you engineer the ecosystem. [See **Niche Construction Theory**]. But don't flatter yourself that by so doing you have somehow opted out of evolution. Earthworms can alter the pH balance of soil all right, but they can't blunt the blackbird's sharp eye. For all their hard work, earthworms barely dent the vast field effect of selection pressures, the hundred thousand shocks that flesh is heir to.

Lichen and moss weather rocks and create soil, making life cushier for subsequent generations of lichen and moss. Does that stop lichen evolution? Is that it for moss? Should they have stuck it out the old way?

Or consider the Texas leaf-cutter ant, *Atta texana*. Secretion from Texas leaf-cutters' huge poison glands kill virtually all bacteria and fungi dead, thus bequeathing even the most delicate ant offspring increased fitness and survival prospects. By doing this, Texas leaf-cutter ants display more not less random mutation. The more variety in the ants that survive to sexual maturity the richer and more diverse the population. So making things cushier for ants far from applying a brake to evolution, widens the scope of natural selection.

Same with us: our hospitals broaden and deepen the human gene pool. And if the wards could just get a decent infestation of Texas leaf-cutter ants, then goodnight E-coli and MRSA!

The idea that hospital maternity wards have put the brake on evolution has a pedigree. (I stress that the following views are very far from David Attenborough's.) The nadir of this thinking came not in the 30s, for all its famous panic about degeneration, but in the 1960s in the work of WD Hamilton.

A generation or two after the modern evolutionary synthesis fused Mendelian genetics with Darwinian selection, WD Hamilton helped develop the gene's eye view of evolution, adding his own Selfish Herd Theory. Famously his work on what he called gene selfishness was popularised in Dawkins' *The Selfish Gene.*

Now, where you and I mourn hospital closures, what Bill Hamilton mourned was hospital openings, because every new hospital was another nail in the coffin of the human race:

'By our acceptance of the Mephistophelean gifts of modern medicine,' he wrote in his 1985 paper *The Hospitals Are Coming,* 'we are visiting on our descendants a disaster equivalent to an asteroid impact [...] A century and a half ago sickly infants died, but now almost any breathing human matter can be perfused and kept alive.'

The unnatural preservation of those he calls 'the ugly and the sickly' is like an asteroid impact in the terrible burden it imposes on those he calls 'the beautiful and the healthy'.

A century and a half ago, poor children in Britain died – as they die now in the Global South – simply because they couldn't get anywhere near clean water, coal fires, regular meals and proper medicine. To survive even to the age of nine probably required more wit, strength, courage and stamina than most of us will ever draw on in decades.

Nor was it just the children of the poor who died. A century and a half ago, Darwin and Wallace buried three children between them. In 1862, Abraham Lincoln's son Willie died. 'A counterpart of his father save that he was handsome,' Willie Wallace Lincoln died not because of defective genes but because of the proximity of the White House to the malarial Potomac River, which the Union troops encamped nearby used as a latrine.

It is instructive how the concept of an outside world – an army, a war, an open sewer beside a malarial swamp – totally escapes Hamilton, as it does so many of the proponents of genocentric theory, with their fantasy scenarios of immortal genes 'sealed off from the outside world.' Nothing daunted, Hamilton extends his formidable grasp of history into pre-history. Healthcare, he decides, is a modern error quite outside the ancient path of human evolution. Recent findings in paleopathology, however, do not support his tough talking.

In 2009, the journal *Anthropological Science* published a paper about a 3500 – 3800 year-old Vietnamese burial site called Man Bac, 15 miles inland from the Gulf of Tonkin.[20] Among the dozen bodies lay one known as Burial 9, whose skeleton shows he suffered from Klippel-Feil syndrome. Paralysed from the waist down, with little use of his arms and restricted neck movement, he would have needed lifelong help to feed and clean himself. In the Man Bac burial site he is buried with equal status to the other dead males of the tribe. No-one in the tribe saw Burial 9 as just 'any breathing matter to be perfused and kept alive'.

'The provision of healthcare,' says one of the Man Bac paper's authors, 'may therefore reflect one of the most fundamental aspects of [human] culture.'

What if we go back further? What about 10,000 years ago?

In Calabria, Italy, archaeologists unearthed a 10,000 year-old skeleton known as Romito 2, who had acromesomelic dysplasia. According to *Nature,* Romito 2 is not only the earliest known case of dwarfism in the human record, but also 'provides evidence of care in the Upper Paleolithic.' But I don't think Romito's bones point to evidence of care, so much as evidence of consensus decision-taking in the Upper Paleolithic. Romito 2's height was about 110 cm, so he couldn't run as fast as his tribe of nomadic hunter-gatherers, when fleeing predators, hunting prey or simply traveling from one bivvy to another. Romito's tribe must have made a collective decision to change their ways so he could keep up. Perhaps they did this from a sense that the tribe couldn't be who they were without him there too. Perhaps because of emotional ties. Or possibly for reasons to do with technique, innovation, and invention. Here's my theory...

20 MF Oxenham, L Tilley and H Matsumura, *Paralysis and severe disability requiring intensive care in Neolithic Asia,* Anthropological Science (2009).

I believe Romito was, if not the human who invented language, then the Italian who taught the Italians Italian. Romito's elbows were fused which limited the mobility of his arms. Now for an Italian, not being able to gesticulate wildly counts as a speech impediment.[21] Pre-Romito Upper Paleolithic Calabrese would have been limited to the odd *cazzo* and *fa vangool* with a lot of extravagant arm action. Romito, meanwhile, takes the sow's ear of stunted expletives and turns it into the sexy silk purse of:

Non lo sai tu che la nostra anima e composta di armonia![22] This may explain why every bone in Romito's body appears to have been smashed into tiny pieces shortly before his death. No-one likes a smarty pants.

Fitness simply means fit for your local environment, and if that environment includes healthcare then you fit. If you are born with acromesomelic dysplasia into a Calabrian tribe who value both you and the virtue of solidarity – you fit. Your inclusion increases the fitnesses of the tribe. This is also true even if the 'tribe' in question happens to be the United States of America in the mid-twentieth century.

In his unprecedented four terms of office, 32nd President of the United States of America, Franklin Delano Roosevelt saved his tribe from its dustbowl depression just in time to save the world from fascism by winning the Second World War. According to biographer Doris Kearns Goodwin, FDR achieved all this not in spite of, but because of the poliomyelitis which left his legs as paralysed as those of Burial 9.[23] She argues that the polio he contracted in 1921 when he was 39 on Campobello Island, New Brunswick, not only broadened and deepened him as a human being, but crucially gave his long White House tenure a crucial blessing in disguise: the rare gift of isolation.

21 My mother's maiden name was Bertani. My grandfather came from the region of Italy called New Jersey.

22 Leonardo da Vinci.

23 Doris Kearns Goodwin, *No Ordinary Time: Franklin & Eleanor Roosevelt - The Home Front in World War II,* Simon & Schuster (1995).

In the days before disability ramps and accessibility statutes, it was a palaver for even the most powerful human being on earth to hit the supper circuit. FDR had a cast-iron excuse for not attending all those Washington functions by which corporations have captured every other White House incumbent. The trade associations, business councils, and banking institutes weren't able to have a quiet word in his ear, or lay a confiding hand on his arm. FDR was the only president who ever got to choose who he hung out with. Uniquely, his downtime was his own.

And so after an evening spent in his White House ground floor study, arranging his stamp collection, mixing cocktails for old friends, or shagging the exiled Princess Martha of Norway, next morning he picked up where he left off with his genial socialism. Sitting up in bed with his breakfast tray on his lap, he pondered:

'What shall I do today? Why I do believe I'll propose that no American citizen ought to have a net income of more than $25,000 dollars a year. Then I'll set up a Civilian Conservation Corp to create one million full-time conservation jobs for 18-25 years olds. For in times of economic hardship, environmental degradation is a luxury we can't afford.'

FDR's Works Progress Administration fostered the same mutual aid we see preserved in the Man Bac burial site, south of Hanoi. This ethic is sometimes lost sight of, however. After I did a gig in North Carolina, I was listening to a late-night radio phone-in show when a woman called up and said:

'I got all these Vietnamese moved into my neighbourhood. I don't like it. Too many of 'em. All these Vietnamese coming over here all a' time. How would they feel – how would they feel if a whole bunch of Americans just decided to move into Vietnam?'

❧ HUMAN GENOME PROJECT ❧

After the Human Genome Project was given $3 billion dollars funding, science journals came out with startling pronouncements that the Human Genome Project was on the cusp of identifying hundreds of thousands of previously undiscovered human genes, including a jealousy gene, a getting into debt gene, a low-voter turn out gene, and a homelessness gene.

The homelessness gene was supposed to be late onset. This worried a lot of people. You might not know you were a carrier until late middle-age. You could be living in the suburban stockbroker belt, never knowing when it might strike.

TREVOR HOWARD: Darling, I can't help noticing that you've been collecting a lot of cardboard boxes lately.

CELIA JOHNSON: Yes, well they look so cosy, especially with winter coming on, don't you know? Silly really, here we are in a six-bedroom mansion, and we've just had a marvelous supper.

TREVOR: And washed down with a rather nice claret. I think I'm going to help myself to a little more.

CELIA: Come near my bottle I'll feckin' snap your bleedin' neck like a twig! Aarrgh unghh! I'll fucking kill you, aaguggghh I got a knife! Sorry darling, I don't know what came over me.

TREVOR: I hope you don't mind me mentioning this, darling, but I can't help noticing that you appear to have lost one of your front teeth.

CELIA: Oh, you noticed. That's very sweet of you. It's silly really. I was in the village and I decided to pop into the chemists on the High Street for my methadone scrip. Now usually I drink it there and then, but Mrs Frobisher, the Pharmacist, is the most frightful chatterbox, and it was such a lovely day so I took the little paper cup outside. Now I don't if you know Mad Angus, I'm standing on the pavement and he says I owe him my methadone. Well, I say to him, quite how do you figure that one out, Mad Angus? Pray tell. And he said that last week he'd give me his bottle of meths. And I said to him, it hurts a girl's pride to have to remind you of this fact, Mad Angus, but what you're forgetting, it seems to me, is that I gave you a blowjob for that meths. To which he replied, No, you gave me the blowjob for those almost brand-new trainers I found on the dead feller.

Ah, I thought, he's got me there! But I suspect cussedness got the better of me, and I necked the methadone, we fell to blows and hence the tooth, or lack thereof.

Of course, what the Human Genome Project actually discovered is that there are in fact only around 25,000 human genes all told, and that no single gene exists that codes for behaviour one-to-one in that reductive, simplistic way in which everyone was talking about at the turn of the century. But by then it was too late. The voodoo mythology had entered the language. Do you have the hard-work gene? The go-getter gene?

I

❧ THE INVINCIBLE IGNORANCE ❧ OF STEVEN PINKER

In *The Better Angels Of Our Nature,* Steven Pinker argues that humans are hardwired for war. Our basic instinct is the killer instinct.

'Chronic raiding and feuding [...] characterized life in a state of nature,' claims Pinker. So much so that fully 15% of prehistoric populations were slaughtered in wars. If you have any doubt about that, he argues, then just look at the state of continual tribal war in which today's last remaining hunter-gatherer tribes live.

To show how 'shockingly violent' today's South American indigenous people are he uses stats from a paper by economists Bowles and Gintis (2011, 2009). Their paper reheats eccentric data from anthropologists Hurtado and Hill (1996), who studied inter-tribal warfare among contemporary hunter-gatherers, including the Cuiva Indians (pronounced – and sometimes spelt – Cuiba).

Okay, so Steven Pinker doesn't use primary sources. That's because he is reviewing all the published evidence on this subject, collating what other researchers have found, and drawing conclusions. You may not like these conclusions, Newman, they may upset your rosy picture of the world, but they are based on a very broad base of evidence. So what does it matter if he doesn't use primary sources?

It matters when innocent victims (including seven children) enter Steven Pinker's bar graph as murderers. It matters when the victims of what Colombian state investigators called 'a genocide against the Cuiva Indians' end up as academic proof of Cuiva violence, because as anthropologist Douglas Fry points out:

'The so-called war deaths are actually indigenous people being gunned down by local ranchers.'[24]

Hill and Hurtado admit as much themselves:

'Members of the study population described here were victims of the "Rubiera massacre" carried out on a Colombian ranch on Christmas Day, 1968, which resulted in the deaths of 16 men, women and children (and left only two survivors who were interviewed as part of this... study.)'[25]

What was this massacre?

It took place at a newly-built cattle-ranch called La Rubiera on the Colombian-Venezuelan border, on December 26th 1967[26] (not as Hill and Hurtado say on December 25th 1968)[27] when, as the *New York Times* reported:

'A group of nomadic Cuiba Indians accepted a range boss's invitation to a meal beside a ranch house. As the Cuibas began to eat, cowboys emerged from the ranch house. With guns, machetes, hatchets and clubs they slaughtered at least 16 Cuibas including women and children.'[28]

La Rubiera massacre became international news because of the defendants' successful plea that they didn't know killing Indians was a crime. The banner headline in Spain's daily newspaper *El Tiempo* repeated their claim to an astonished world: 'Yo no sabia que era malo matar indios.' I never knew it was wrong to kill Indians. All six defendants were acquitted thanks to the 'invincible ignorance' defence then available in

24 Douglas Fry (ed.), *War, Peace and Human Nature,* Oxford University Press (2013).

25 Hill, Hurtado & Walker, *High adult mortality among Hiwi hunter-gatherers: Implications for human evolution,* Journal of Human Evolution 52 (2007).

26 *District Judge Eloy Villamazar, Jefe Agrupacion Rurales de Arauca.*

27 Hill & Hurtado's numerical slip here may incidentally explain how their own mortality study managed to calculate the Ache tribe as ten times more violent than any comparable study has ever found! To anthropologists this finding isn't just anomalous, it's like Jessica Ennis-Hill getting ten times as many points as the next heptathlete. You'd demand an enquiry.

28 *Cowboys' Retrial Near In Colombia - First Jury Acquitted Six Men In Massacre of Indians,* New York Times (January 7, 1973).

Colombian law: you can't be convicted if you didn't know that what you did was a crime.

Come the 1973 Bogota retrial, however, prosecutors showed that the cowboys had tried to hide the dead by incinerating the bodies and mixing their bones with cattle bones.

This, the prosecution argued, proved that they well knew they had done a deed for which they would be punished if found out, and therefore contradicted their original plea that they were invincibly ignorant of the fact that killing Indians was a crime. In the same way that these cowboys mixed the bones of their victims in with the bones of cattle, so the massacred Cuivas' bones are mixed into Hill and Hurtado's graphs, pulped into bone meal for Bowles and Gintis, and sprinkled as fertiliser for Pinker's fantasy thesis about the chronic violence of indigenous tribal peoples.[29]

Here below are extracts from the witness statements of the only two survivors of the La Rubiera massacre.[30] This is the first time these parts of their testimony have appeared in English, I believe, because I translated them from the Spanish myself.

TESTIMONIO DEL INIGENA ANTUKO N.

One day in December Marcelino Jiménez came to where us Cuiva Indians were working on the Manguito ranch.[31] He told us that if we went up to La Rubiera ranch house they were going to give us a feast, with panela (sugar loaf)... rice and fresh meat.

After inviting us, Marcelino walked up to La Rubiera, while we went in four canoes. We were Luisito, Chain, Ramoncito, Ceballos, Guafaro, Luisa, Doris, Bengua, Cermelina, Lilia

29 Pinker's rigorous scholarship was highly praised in peer-review journals, where no reviewer blinked an eye at his inversion of the historical record, it being beneath the dignity of dispassionate science to trifle with irritatingly irrelevant mass murder.

30 *Investigacion sobre el genocido de 16 indios Cuivas perpetrado en la region del Capanaparo,* Servicio de Seduridad Rural de Los L.L.O.O., Agrupacion Arauca, Departametno Administrativo de Seguridid, Republica de Colombia.

31 Marcelino del Carmen Jiménez, a nineteen-year-old Colombian ranch hand sent there by Rubiera ranch owner Luis Enrique Morin.

Quintero and me. The children who came were Carmelina (Doris's daughter), Daisi (Dionisia's daughter), and the boys Miye, Arusi, Alberto Santana, Julio Guamare and Isidoro.

TESTIMONIO DEL INDIGENA CEBALLOS N.
We left the canoes on the Capanaparo River, which the Colombians call Black Creek, and went up to La Rubiera. Once we got there we wanted to go straight back, but Marcelino told us that we were expected at the dinner which was a big spread with a punch bowl. Outside the ranch house Luis Enrique Morin, or "Chamua" as he's known, was standing with Helio Torrealbe, known as Julio [...] and when we were eating, they came out of the ranch house with knives, revolvers, a rifle and a shotgun. With them were people we had never seen before who had revolvers in their hands. We all ran, but they had bolted the gate of the corral behind us so we couldn't escape.

[...] They stabbed Doris and her daughter Carmelina against the wall of the house. Ramoncito fell by the barbed wire fence, they killed Luis further away by an oak tree.

TESTIMONIO DEL INIGENA ANTUKO N.
At dusk, me and Ceballos hid in a tree. The next day, we saw them dragging the dead bodies behind a mule [...] That afternoon, we went down to the Capanapora river and each towed an empty canoe back to El Manguito.

[...] My wife, Bengua, and Ceballos's wife, Luisa, were among the dead.

We arrived at Manguito and told everyone what had happened and they told the ranch owner Marcelo Tapias and Father Gonzalo.

The photograph shows Antuko and Ceballos giving their testimony with Spanish-born, Dominican priest Father Gonzalo acting as interpreter, translating the survivors' Guahibo into Spanish.

In *Hunting Indians,* a study of La Rubiera and other massacres in the region, anthropologist Carwil Bjork-James asks the key question that cuts to the heart of everything:

'What ideas made these events possible?'

Ideas like Pinker's make these events possible, because these are very old ideas, which have been around for a long time, but are no less dangerous for that. Bjork-James quotes a frontier government official petitioning to Bogota to send machine guns on finding himself in an 'entirely savage state.' A little later a Colombian navy riverboat machine-gunned Indians on the banks of the Rio Meta. No more savagery in the state now, thanks to those heavy weapons.

For if the 'state of nature' really is savage, if nomadic foragers really are irremediably vicious, then that entails extirpation almost as self-defence. For Pinker, as for the cowboy ranchers who planned their Yuletide slaughter, the innate savagery is a given. The question, then, becomes, do you let it live among you?

You may say:

'Look, Pinker is dealing with the Big Picture and you are

pricking holes in the canvas when really you should step back and look at the vast sweep of the epic picture he paints from prehistory to now.'

Okay, then, let's look at the prehistory part of his argument.

'Chronic raiding and feuding [...] characterized life in a state of nature,' argues Pinker. To prove that per capita prehistoric war deaths were higher than even the bloodiest conflicts today, Pinker cites 21 archeological sites from all around the world, each and every one a murderous moshpit of massacre and megadeath.

Archaeologist R Brian Ferguson has painstakingly reviewed the evidence.[32] Among Pinker's 21 prehistoric sites, Ferguson finds three sites – located in what would now be Algeria, Sudan and India – in each of which the total number of war dead is one. If only modern wars had that kind of body count! Imagine if World War 1 and World War 2 got their names because one person was killed in World War 1 and two in World War 2. I don't know about you, but I'm starting to feel rather puckish about our chances of surviving World War 3.

Ferguson finds that Pinker's list includes four cases of the same site being counted twice, and a site at Gobero, Niger – where there are *no* war deaths, or even any recorded homicides *at all*. This is a 16,000-year-old site and so perhaps the absence of any homicide victims may be put down to the fact that we hadn't yet found a way of killing each other:

'I'm telling yer, if I ever get this blowpipe to work, you're fucking dead!'

'Yeah? Well, you better be quick, 'cos I'm only a few years from getting this hafted axe to stay on the handle.'

At last, Pinker lights upon a South Dakota archaeological site that really does present evidence of actual prehistoric mass slaughter. The only problem is that it's from a battle that took place in 1325 AD.

32 R Brian Ferguson, *Pinker's List - Exaggerating Prehistoric Wartime Mortality*, in *War, Peace and Human Nature*, Edited by Douglas Fry, Oxford University Press (2013).

Now, in 1326 AD, just to give some context, Clare College, Cambridge and Oriel College, Oxford admitted their first undergraduates. What most people mean by prehistoric doesn't include teams that have been on *University Challenge.* I don't want to be New World-ist, and I respect that in archaeological terms, prehistoric sometimes means before a local written record, but still, for most people prehistoric doesn't begin twenty years after Robert the Bruce is crowned King of Scots, (following his successful mastodon charge at the Battle of Bannockburn), and one thousand years after historians such as Tacitus and Livy wrote accounts of Roman military campaigns.

...Speaking of Romans, there's a curious similarity between the argument of Pinker's *Better Angels* and the Roman poet Horace's theory of evolution:

> *When living beings first crawled on earth's surface, a dumb and filthy herd, they fought for acorns and lurking places with their nails and fists, then with clubs, and so from weapon to weapon [...] until they discovered language, by which to make sounds and express feelings. From that moment they began to give up war, to build cities and to frame laws that none should thieve or rob or commit adultery.*[33]

As the rich get richer, I suppose their opinions, like their underfloor heating, will become more Roman. But Horace lacked the archaeological evidence and forensic techniques which enable us to draw different, and less gloomy, conclusions today:

'The worldwide archaeological evidence,' says Douglas Fry, 'shows that war was simply absent over the vast majority of human existence.' Reviewing Fry and Ferguson's findings, Noam Chomsky concurs:

'War is a recent innovation in human history, not an ineradicable curse.'

33 Horace, *The Satires,* Book I, Satire III (~35 BC).

Despite the above caveats, however, I confess that Steven Pinker has indeed challenged some of my most deeply held beliefs. If I have been hard on some aspects of his argument it is a defensive reaction. For truth be told, Steven Pinker has forced me to face something I don't like to think about: the human propensity to perpetrate the most shocking and deplorable acts. He leaves me with a series of questions, each more depressing than the last...

Are we hardwired by evolution to twist and manipulate data? Is Steven Pinker simply helpless to resist a genetic imperative to juke the stats? Has he no free will? What can explain the fact that most people most of the time don't juke the stats? Is it just luck that the better angels of our nature sometimes intervene and sometimes don't?

J

❧ JUST SO STORIES ❧

For generation after generation, children of all ages have read Rudyard Kipling's *Just So Stories* and wondered why they're not a patch on *The Jungle Book*.

In such stories as *How The Leopard Got Its Spots* and *How The Camel Got His Hump,* Kipling starts with the trait and then works back to an ingenious explanation for each morphological idiosyncracy. Many gene-centred accounts of evolution are notoriously just such *Just So Stories.*

A few years ago, for example, BBC national news gave respectful airtime to an evolutionary biologist who had discovered that the reason our fingers and toes corrugate in water is a cunning adaptation designed to increase our grip when climbing out of Pleistocene rock pools.

The enhanced grip of wet feet will be news to all the A&E nurses who each day take delivery of concussed and bloody nudes shrink-wrapped in their own shower curtain.

If extinct hominids lacked rippled fingertips and toes, perhaps that explains why so many of them are found in sediment at the bottom of dried up rock pools. Seems they just couldn't clamber out!

You'd think at least one peer-reviewer, or maybe a passing five-year-old, might have told the scientist about that time they left a book out in the rain and the pages went all ripply. But no newsroom wants to run a story about well-known properties of water. They want to feature cutting-edge science about genes, and the great mystery of rippled fingertips explained.

This example is good clean fun and harmless nonsense, but many evolutionary Just So Stories strive to make the way we live now seem right and natural by extrapolating profound and ancient rationales for recent quirks. (See, for example, the **Evolutionary Psychology** section for a magnificently specious explanation of why girls like pink and boys like blue).

In and of itself, though, there's nothing so very wrong with an evolutionary Just So Story. In fact, it can often be the very type of speculation that opens up vast expanses of intriguing inquiry. Here's an example of the sort of thing I mean. I guess Kipling would call it *How Humans Got The Whites Of Their Eyes*. The story explains why alone among primates humans have eye-whites, or sclera – just like dogs.

One hundred thousand years ago, the population of Europe was entirely Neanderthal. Then about 80,000 years ago, anatomically modern humans came out of Africa with tame wolves or African hunting dogs. These dog packs may have helped us eat on the move. So far I am 100% in agreement with the story, and only start to demur when it gets to the part about how owner started to look like dog the further we traveled together as nature selected for humans with dog-like eyes.

What evolutionary benefits might come from having dog-like eyes with white bits in them? Well, if a dog can look into your eyes and think, 'Do you know, I actually believe he's trying to tell me something,' then that dog is more likely to risk its life to save you from *Dino felis,* the giant prehistoric jaguar that preyed on dog and human alike. And that's how the humans got the whites of their eyes. The End.

It's a good 'un, isn't it? But to share a trait doesn't entail co-evolution. This is as true of sclera as it is of other traits we share with canids. African hunting dogs, for example, divide labour and are fairly non-hierarchical.[34] Unlike some ethologists, however, I don't think we learnt our sense of equality from

34 Wolfdietrich Kuhme, *Communal Food Distribution and Division of Labour Among African Hunting Dogs,* Nature (1965).

dogs. I think we already had our own. Of course, to have things in common helps form a bond. If we both rooted for the underdog then that might have helped us see eye to eye.

Incidentally, the phrase 'rooting for the underdog' tells you everything you need to know about British hypocrisy. Our British sense of fair play, we like to say, is summed up by how we always root for the underdog.

'Yes, sir, that's us. Last night at the ten pound a ticket organised dogfight in the pub cellar, we cheered the game little one-eyed Jack Russell (3-1 odds) with his ear hanging off who put up such a gallant show against the bull mastiff (11-10 against). That little Jack Russell was still fighting even when he was having to drag his broken back legs behind him. Go on, Jackie, we shouted right up until the mastiff got into his giblets, and his owner had to shoot him with a 12 bore. Splendid stuff! Applause all around, and even a dabbing of sentimental eye with kerchief. Now, your people from other countries? Do they even have dog-fight etiquette?'

K

KIN

'Family feelings are designed to help our genes replicate themselves.'[35]

For Steven Pinker a Trickster Nature cons and hoodwinks us with *Little House On The Prairie* sentiments into spreading her nefarious molecules. But why does Nature have to trick us out of our own nature? There's no reason, except that Steven Pinker has fallen into the fallacy acutely observed by the philosopher AN Whitehead a century ago:

> *Scientific reasoning is completely dominated by the presupposition that mental functionings are not properly part of nature.*

Darwin deplored the fallacy of a true self separate from our evolved self. For him, as for Whitehead, mental functionings are inextricably part of nature. But not for Steven Pinker:

'And if my genes don't like it,' he writes in *How The Mind Works*, 'they can go jump in the lake.'

Steven Pinker's view of humanity derives, then, from Darwin's contemporary the famous Victorian scientist, Dr Jekyll. The more nineteenth century the sentiment, the more twenty-first century the language. Here's that quote in full with which I started this section:

'Family feelings are designed to help our genes replicate themselves. Our emotions about kin use a kind of inverse genetics to guess which of the organisms we interact with are likely to share our genes.'

35 Steven Pinker, *How The Mind Works*, Penguin (1999).

I don't know what the word 'organism' has ever done to Steven Pinker to make him treat it so badly. Since he starts off the sentence talking about human emotions, then 'people' is the natural choice of word here. But Steven Pinker can't be using folk terms like 'people'. So he goes for what would be a more scientific word, if only science meant syntactical incoherence, for to say organisms in this context makes no sense. Nobody uses any kind of inverse genetics to guess whether the snail they are flicking over the garden wall shares their genes. You use your loaf. Guessing doesn't come into it:

'Hang on a sec, didn't I meet this snail at my Cousin Flo's wedding?'

Now, *people*, on the other hand, might well try to guess if they are related to other people. So the word 'people' would make sense. But some things are more important than making sense. Making an impression, for example. The word 'organisms' makes a very good impression. It says Dispassionate Scientist At Work. And that's pretty important to flag up right now because we are entering the Rocky Realm of Cold Hard Truths. Some of you may want to turn back now. The rest of you – buckle up for some heavy shit. Ready? Turns out all those warm soft feelings we cherish are illusions. Only those who share your genes really matter to you when all is said and done. Yeah. Heavy. Can you handle the ugly reality that family feelings are little more than kin-selection and less than kind? Can you live with the dark side?

Behind this macho posturing, however, lies a peculiarly cosy view of family life that averts its modest gaze from anything unsightly. Not so Pulitzer-Prize-winning author Marilynne Robinson, who in her critique of Pinkerism asks:

'Might there not be fewer of these interfamilial crimes, honour killings [and] child abandonments,' if family feelings really were designed to help our genes replicate themselves?[36] And if 'our emotions about kin' are calculating 'which

36 Marilynne Robinson, *Absence of Mind*, Yale University Press (2011).

organisms share our genes' then why, asks Hilary Rose, did slave owners in the *antebellum* South, routinely sell children fathered upon slaves without a backward glance?[37]

So much for genes getting us to dance to their music.

KROPOTKIN

At the end of the nineteenth century, Peter Kropotkin came back from the brink of death to make a major contribution to evolutionary theory, with his book *Mutual Aid: A Factor In Evolution.*

I first learnt about Kropotkin from Dr Natasha Rodionova, of the University of Novosibirsk. In a remote Siberian station one night, its small coal-fire insufficient to stop our breath steaming like the samovar of strong tea on the table, Natasha told me his story.

'In late-nineteenth-century St Petersburg,' she said, 'Prince Peter Kropotkin was leading a precarious double-life. By day, an internationally renowned scientist, by night part of a clandestine organisation bent on the overthrow of the Tsar. When Kropotkin was told that the secret police, the dreaded Third Department, were onto him, he didn't flee the country, but stayed to deliver lecture on glaciation to the Russian Geographical Society.'

'Sounds risky,' I said.

'Risky. What do you know of risky? Once I was working in Kremlin – don't ask me how I came to be there – I was cultural attaché and it fell to me to welcome an official delegation of the Chinese Communist Party led by none other than Chairman Mao himself, who was wearing a blue tunic with green trousers. In an effort to break the ice on what was a rather stuffy, formal diplomatic occasion, "Chairman Mao," I said to

37 Hilary Rose, *Alas Poor Darwin, Arguments Against Evolutionary Psychology,* Edited by Hilary Rose and Steven Rose, Vintage (2001).

him, "blue and green should never be seen except upon an Irish Queen." He looked at me and went, "Ha, ha. Her whole family and her village."'

'Why doesn't Kropotkin flee?'

'He does not flee,' she said, 'because he believes nobody will ever see through his cunning disguise. When Kropotkin speaks to hundreds of revolutionary artisans at the underground meetings he assumes an alter ego. He pretends to be a weaver called Borodin. No-one is supposed to know that Borodin the humble weaver is actually Prince Peter Kropotkin, the celebrated aristocratic scientist. But I fear he is not quite as good at the double-life as he imagines, his imposture was not perhaps so convincing. It rather seems he used to come straight from Imperial Court functions at the Winter Palace, change out of his princely robes, dress up as Borodin the weaver, appear at a clandestine revolutionary meeting, stand on a platform in front of hundreds of workers and say: "Cor blimey, mateys! What about these lah-de-dahs toffs swanking around in their fancy coach and horses, eh? Me? I'm happy wiff the ole donkey 'n' cart, long as I can have a knees-up wiff a pint o' beer in me hand."[38]

'Kropotkin is thrown into the notorious Tsarist dungeon the *Petropavlovskaya Krepost*, condemned to solitary confinement for the rest of his life. His cell is so damp that after two years he is given two months to live.'

'Desperate,' I said.

'What do you know of desperate, *golubchik,* you from the pampered West, with your crochet lady toilet roll covers? There is a bucket in the corner of Kropotkin's cell for his kaka. But there is no crochet lady on the bucket. If there was, she would be filthy. You would not be calling her crochet lady, but crochet hag, but in her heart she is still lady.'

'Who are we talking about here?'

38 I don't mean to offend any Paul Weller fans here. But for neither man is the pretence successful. Only the consequences are much worse for Kropotkin.

'Do not try to psychoanalyse me. In my village we have an old saying: Help, help the Cossacks have burnt down our farm and killed everyone but me! Ah. Tea is ready. Shall I be mother?'

'Yes, please.'

'OK,' she said and shouted: 'I have no idea who your father was!' Once the tea was poured, she continued. 'After two years in solitary, Kropotkin is given two months to live and moved to the prison's hospital wing to die. But here the cell is dry, there is a fireplace, Kropotkin's health revives, and each morning he finds he can shuffle a little further across the hospital exercise yard, where he notices that the courtyard gates are opened each morning for the delivery of timber for a new TB ward. Thus plans are hatched for what turns out to be the most celebrated escape in all Europe, an escape plan as elaborately choreographed and with almost as large a cast as a dance sequence from *Oliver*! Worked out by his sister, the escape plan involves a violinist playing a mazurka at an open window across the street, a woman carrying a bunch of red balloons which can be seen over the prison wall, a relay of people eating cherries on street corners in a one mile radius of the prison, a racehorse put between the shafts of a carriage. In fact this champion racehorse proves to be too fast: when Kropotkin makes his dash through the gates and jumps into the carriage, the racehorse takes the first corner at such a clip that the carriage tips onto one wheel and its roof gets wedged into a narrow alley. Kropotkin is stamping the carriage floor down flat, soldiers are running towards him, and the whole escape might have ended there and then, but for the sudden appearance of five hundred dancing cockneys singing *Who Will Buy My Lovely Red Roses?*'

'Perhaps,' I suggested to Natasha, 'Kropotkin didn't flee because he really really wanted to give this talk on glaciation?'

'Glaciation,' she repeated witheringly. 'What do you know of glaciation? When you have seen the only man you ever loved swept away by a glacier *then* you can talk to me about glaciation. For me the U in U-shaped valley spells unutterable loss, untimely death, ululating unguish.'

'Forgive me. Natasha Dmitreya.'

'It is hard for you to imagine now, but glaciation was once a revolutionary idea. If valleys and rocks and mountains can be swept away then why not the Tsar? Here I have a gift for you.'

Natasha's gift was a thing of beauty: a signed first edition of Kropotkin's autobiography, *Memoirs of a Revolutionist.*

In this wonderful memoir, Kropotkin describes how he made good his escape:

> *I crossed Sweden without stopping anywhere, and went aboard a steamer at Christiana. As I went to the steamer I asked myself with anxiety, 'Under which flag does she sail, – Norweigian, German, English?' Then I saw floating above the stern the union jack, – the flag under which so many refugees, Russian, Italian, French, Hungarian and of all nations, have found an asylum. I greeted that flag from the depth of my heart.*

Anglophile though he was, Kropotkin found England and in particular London a terrible let down. Now, I sometimes feel I should have been born a Russian. I feel they understand me better than my own people. (At the moment, for example I am reading Tolstoy in the original Russian. Just the verbs). Other times, though, I think I'll never understand the Russians. And this next extract from Kropotkin's memoirs definitely falls into the latter case:

> *I confess that at first I found London to be a disappointment following the intoxicating cultural, political and artistic ferment of Continental Europe. But then I discovered Bromley.*

While Kropotkin is in Bromley (it may not be Tashkent but it is Kent) he reads T H Huxley's splenetic article *The Struggle for Existence in Human Society* (1888) in which Huxley says that nature is a gladiatorial contest in which the weak go to the wall, and woe betide us if we fail to apply this maxim to human society for we shall sink under the burden of supporting those unfit to live. This appalling pamphlet inspires Kropotkin to write *Mutual Aid – A Factor in Evolution.*

Yes, there's a struggle for existence in nature, he argues in *Mutual Aid*, but it is most often a struggle against the elements, against a hostile environment, in which struggle those species that stick together do rather better than ones that don't. It's penguins plural that have evolved to withstand six week long Antarctic blizzards by huddling together and sharing incubation duties. Penguin*s*. The penguin – singular – hasn't evolved to do anything except menace the good people of Gotham City.

If selfishness is indeed a fundamental principle of nature, as claimed by T H Huxley and Herbert Spencer (the man who coined the phrase 'survival of the fittest') then how come when vampire bats return from a hunting expedition successful hunters feed unsuccessful bats with blood from their own mouths? Why does the vervet monkey risk her life to warn the rest of the troop that a leopard's nearby?

Kropotkin described how sentinel crabs in Brighton Aquarium guard the soft-bodied crabs during molting season, and will die protecting them. And he looked at the way in which female African buffalo appear to *vote* which way the herd should move. During rest time the cows all lie down facing different directions. At the end of the rest time, whichever cow has accumulated the most bodies behind her, that's the direction the whole herd moves. And as they migrate across the Serengeti, should one buffalo fall into a ravine, not only does the rest of the herd make strenuous efforts to rescue her but they also, somehow, resist the urge to say:

'And it was your idea to come this way and all!'

When he extends his argument to human beings, Kropotkin has a star witness in Charles Darwin, who, in 1871's *Descent of Man* says:

> *Those communities which included the greatest number of the most sympathetic members would flourish best and produce the most offspring.*

Nothing could be truer and nothing could be further from current understanding of evolution by natural selection. Having actually been a soldier, however, Kropotkin is less keen on putting this into fighting talk than Darwin is here:

> *[A tribe] [...] always ready to aid one another, and to sacrifice themselves for the common good [...] would on the whole be victorious over most other tribes and this would be natural selection.*

For Kropotkin, the selective advantages accrue not so much because of platoon morale, as Darwin imagines, but because a diverse population is in better fettle to meet the staggering diversity of shocks and challenges with a diversity of responses:

> *When humanity makes progress it's nearly always because of nervous wrecks, invalids, weaklings, the chronically ill and infirm and so-called inferior people.*

Darwin himself fits the bill here by having been severely debilitated for the best years of his life by Chagas disease.

Now, to be clear, just so there are no misunderstandings, what no-one is saying, not Darwin, nor Kropotkin, nor even the Prince Crackpot writing these words, no-one is saying that competition isn't as intrinsic to life as interdependence. It's just that competition doesn't have what it takes to drive evolution very far. That's all.

When I asked Natasha why negative, pessimistic views of nature are seen as somehow more scientifically rigorous than sunny ones, she replied:

'Despair is its own sentimentality. I remember my mother used to say to me, "Natasha, medicine only works if it tastes so foul as to make you feel physically sick." But now that she is on a morphine drip the views have changed.'

L

❧ LAMARCK ❧

Biology students learn about Jean-Baptiste Lamarck as a kind of comedy overture to Darwin. The joke's on him for theorising how the physical traits you pick up in your life are passed from parent to offspring. Everyone has a good old laugh off him saying the giraffe gets its long neck from stretching up into the trees, or that the blacksmith's son inherits his dad's big biceps. Crazy Frenchman! What endearingly silly Just So Stories people believed in before Darwin turned up and knocked Lamarck's cockamamie ideas into the dustbin of history.

Except it just so happens that Darwin went to his grave believing in the inheritance of acquired characteristics. He thought it was a weaker force in evolution than other factors such as natural selection, sexual selection, or the direct action of the environment, but still very much a force to be reckoned with. More than a decade after *Origin of Species*, Darwin argued that:

> *'Gestures and expressions become hereditary through long practice.'*

And:

> *'actions which were at first voluntary, soon become habitual and at last hereditary.'*

Both quotes come *The Expression of the Emotions In Man and Animals.* The list of illustrations to this book is brilliant, by the way. This is the index: Photograph of insane woman, Hen driving dog away from chickens, Swan driving away an intruder, Small dog watching a cat on a table. Idiots weeping. Chimpanzee disappointed and sulky – it's hard not to feel that we are reading Darwin's YouTube browsing data. All that's missing is George Eliot FAIL!

To espouse even a modified idea of neo-Lamarckism is to run the risk of looking silly. This sort of fear explains why scientists are often the very last to take on board what the science shows, to properly integrate it with their picture of the world. Galileo, for example, refused to accept Kepler's detailed diagrams and deductive proofs demonstrating the eliptical orbit of planets. In a harmonious rational efficient universe, thought Galileo, orbits simply must be circular. They have to be, otherwise you'd have chaos and a Labour government. A pre-emptive philosophical commitment and fear of looking silly blinded him to Kepler's light. In the same way, evolutionary biology prefers to look the other way rather than to properly internalise the new insights into the ways in which the inheritance of acquired characteristics has been put back in play by new discoveries in epigenetics and the role of the environment in developmental plasticity.

❧ LANGUAGE ❧

Richard Dawkins has a very idiosyncratic idea about the origin of language. Language evolves, he contends, as a way of detecting cheats and as enhanced strategy of counter deception. What a miserable Scrooge-like view of the word!

NEIGHBOUR: Hello Richard! Glorious morning, isn't it?

DAWKINS: Is it money you're after?

Now in this book I promise I will, to the best of my ability, focus on the ideas and not make this in any way an ad hominem, personal attack on any of the leading Neo-Darwinist quacks. However, I would at this juncture like to obtrude one biographical datum for it is, I feel, germane, and has to do with an Oxfordshire postal worker friend of mine who, for a time, had Richard Dawkins's Abingdon mansion on his round. Now, my mate worked for Parcelforce, which meant that he had to get Dawkins to come to the door and sign for parcels and packages – a procedure which, my friend says, was never less than fraught. He used to crunch up the gravel drive, ring the doorbell, and Dawkins would come to the door.

POSTMAN: Royal Mail. Can you sign here for the parcel, please?

DAWKINS: So, postman are we? But of course you are. You're wearing a postman's outfit. Red van parked at the foot of the driveway, no doubt with its full complement of black and white cat, I shouldn't wonder. You're a postman with a clipboard and a helpfully supplied biro. All I have to do is sign my name and the parcel's mine! What could be more straightforward than that? Unfortunately, the rules of game theory hold that the opening proposition is always a gambit or ploy. So what, I wonder, is really your game, eh, postie?

POSTMAN: Can you just sign for the parcel, please? I've got a lot of other deliveries to make.

DAWKINS: *Whoah up! Wait a minute, Mr Postman, whoah, whoah, whoah, whoah, Mr Poh-woah-ohhhstman.* This curious urgency of yours makes me suspect that you hoped to find me out, my wife in and receptive to your advances, so that you could deposit not a parcel into my hand but spermatozoa into her fallopian tube. I then raise the concomitant progeny in the delusion that they are my genetic material, leaving you, free from the burden of childcare, to putter around in your red van impregnating the good wives of the parish.

POSTMAN: Can... you... just... sign... for... the... bloody... parcel.

DAWKINS: Ooh-hoo! The mask is slipping! The thin carapace of civilised man falls away to reveal the Stone Age killer beneath! Well, if that's the way you want it, let us strip naked and wrestle!

And Dawkins used to strip off all his clothes and then piss in a circle on the gravel drive. And he'd make my mate strip off all his clothes and join him in this steaming urine arena. And the two of them would wrestle naked back and forth across the gravel. Now, if my mate won the wrestling match, Dawkins would accept the parcel. If Dawkins won, my mate used to have to leave one of those chits that say, We called and you were out.

Now as you can well imagine, a man like Dawkins gets – what? fifteen, twenty parcels a week, books to review, all that, and my friend was determined not to keep losing consecutive wrestling matches, so after a while he developed his own strategy, his own tactics. He'd crunch up the gravel drive, but before he'd even rung the doorbell, he'd strip naked, piss his own circle on the gravel drive to give him a territorial advantage, smear mud and gravel over his face and chest, and then psych himself up with an improvised haka. And only then ring the door bell. Of course, one day Dawkins' mum answers the door, doesn't she? My mate tries to style it out.

'We warned you this would happen with privatisation.' Complaint goes into head office. Guess whose side of the story management believe?

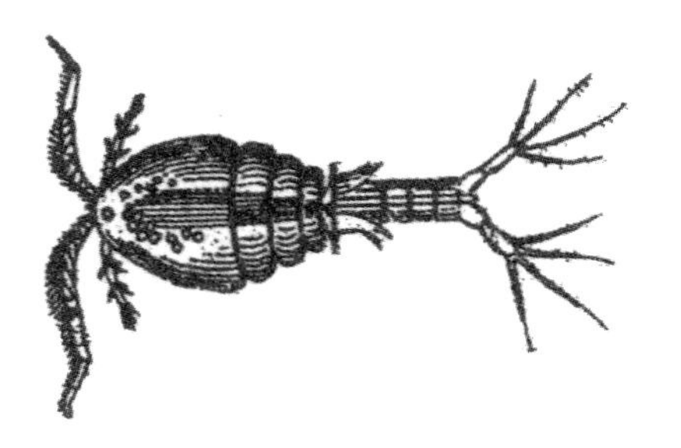

M

❧ MIGRATION ❧

Once upon a time, any greylag geese who were late to arrive in Baffin Bay would miss out on the plum nesting sites. Perched on the edge of a rocky precipice, exposed to predators and Arctic winds, they'd be the first geese to die. Not anymore. Since Baffin Bay spring temperatures rose by 4°C, a little late is very good for the goose. It is the punctual geese who are being eaten by bears. Encouraged by early snow melt to come inland, polar bears wander over turf and tundra scoffing tens of thousands of freshly laid goose eggs. Geese who get a bit lost over Vancouver, on the other hand, who don't rock up in Baffin Bay until the last bears are trudging home to the ice and snow, sick to the back teeth of raw eggs, have never done so well. Disrupted climate cues may therefore be selecting for klutzy geese.

In West Greenland, climate change appears to be leading to a rise in gormless reindeer, whose numbers are overtaking resourceful ones. Intrepid reindeer who let nothing stand in the way of their migration to West Greenland will battle through snow and ice, rain, wind and wolves only to find that upon arrival at the overwintering ground, the forage has not yet appeared, and so they starve. Moron reindeers, on the other hand, who keep stopping for fag-breaks or to ask passers-by for directions, will arrive just as the forage is coming into leaf.[39]

39 Robert Newman, *Disrupted Climate Cues Lead to Rise In Moron Reindeers That Are Like Total Divvies*, rejected outright by journals Nature, Enviro Evol, PlosBiol, Physical Geog., and Science.

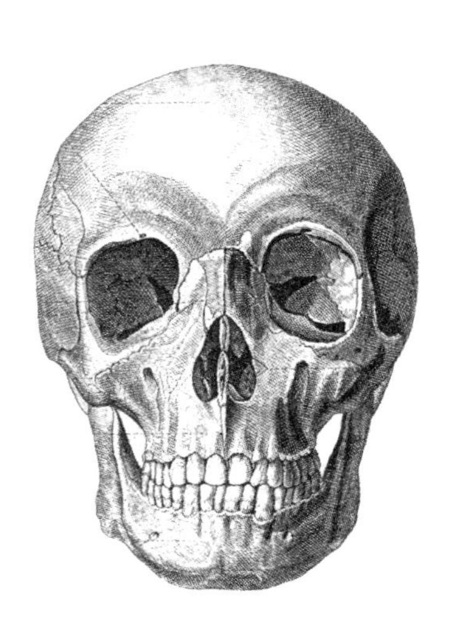

N

NEMATODE

The flatworm nematode can not only be taught its way through a maze, by being given a slight electric shock if it goes down a dark-painted passage, and none if it goes down a light-painted passage, but – and here's the spooky thing – 'the memory appears to reside in a substance'.[40] For if a nematode that's learnt the maze is fed to another one, that nematode knows its way through the maze first time of asking.

Now, there, my friends, is one attribute that I am mightily relieved we humans do not possess. For if we did, you know that clever poor students would be fed to rich thickos. And woe betide any of us who tried to speak out against this. If you went on that show, *Question Time*, where each week topical issues are discussed by a cross-section of millionaires, and spoke against this, they'd denounce you:

'Typical lefty! Taking away opportunity in the name of some specious notion of equality! I actually think it's rather noble that some of the less advantaged State-school students have a chance of attending a first-rate academic institution by being carried there within the lower intestine of a more affluent student. After all, that is why it's called Eton!'

40 David Attenborough, *Life On Earth*, BBC (1979).

❧ NEOTENY ❧

Have you ever noticed how much more like juvenile chimps we look than adult ones?

But there the similarity ends. While the baby chimp grows up into a hairy, long armed chimpanzee, six times as strong as an adult human, we keep the baby face until our dying day. In zoological jargon retaining juvenile features into adulthood is called neoteny. If chimpanzees are *Pan trologdytes*, then we are the Peter Pan primates. Although that said, it's certainly not *me* that spends all day on a tyre swing, lemme tell ya.

Humans prolong development longer than any other primate (and primates longer than any other mammals). If we're the perpetually immature ape, perhaps the first thing we did when we discovered fire was to light our own farts. But, as Alfred Russel Wallace wrote in letter to Darwin, 'there's more to neoteny than being the first species to be able to light your own farts'. What I think he meant was that our arrested development may be the secret of our success.

Prolonging infancy entails years of creative play, of doing things your own way. It means more time going where the grown ups tell you not to go, getting interestingly lost, and stumbling upon something new.

Neoteny may also have helped us realise our latent language capacity. Infants have a wider vocal range, they continuously call for parents, burble experimentally, eliciting an instinct in adults to do a lot of scat-jazz doodles in return. For this reason, I truly believe our first words were probably:

'Zap-bapa boo-boo, diddly woozzee, shabbadoo-coochie-coochie-boobabloo!'

NICHE CONSTRUCTION

Niche Construction Theory is one of the hot new areas in evolutionary theory. This being so, I'm thinking that if I can be the first to sum it up in a neat, pithy epigram, then I'll be quoted in all the literature, invited to conferences, and generally seen as a player, as someone who was in on the ground floor. So here's the list so far.

Niche Construction is how:

- organisms make ecosystems make organisms.[41]
- organisms select environment as much as environment selects organisms.
- what you do to where you live changes your kids.
- organisms create a niche which increases the fitness of new types of offspring unfit for life outside the niche.

Hmmm... Maybe I'd better just give an example of niche construction at work.

In Ecuadorean tropical forests, lemon ants, *myrmelachista schumanii*, create spooky dead zones called Devil's Orchards, so called because nothing else grows but a single species of tree *Duroia hirsuta*. Lemon ants poison with formic acid any other types of tree, bush, flower, or fungus, to create the ugly monoculture they find so beautiful.

It's a Devil's Orchard to us but heaven to them. The lemon

41 A passing similarity with 'Tetley's make teabags make tea' may, however, involve me in a lengthy Intellectual Property Rights court case.

ants have made the world a better place for lemon ants. They have turned a big bad world into a nice little one, where not just the strongest and toughest survive, but the weak, the vulnerable and the fussy as well.

So, lemon ants don't only pass on chromosomes to their oviposited offspring, they also bequeath a local environment. They do not submit to being selected against, but rig selection pressures in favour of the whole colony. And this is an example of Niche Construction.

In case you're wondering why they are called lemon ants, by the way, it's because they taste a bit lemony. Turns out there's someone at the Linnaen Society – the body in charge of taxonomic classification – whose job is ant-taster. Since all ants look pretty similar, the ant-taster has to suck, chew, or lick ants and write up the flavour. Once you know this some of the weirder-sounding ant names make sense. As well as the lemon ant, there's actually a Salt 'n' Vinegar Ant. The Linnaean Society were very pleased when I myself identified a previously undiscovered entomological species, the I Can't Believe It's Not Beetle Ant *(Myrmelachista newmanii).*

Niche construction theory says that an animal doesn't just find an ecological niche, it makes one. And not just animals, trees too.

In *Niche Construction – The Neglected Process In Evolution,* the authors describe how some indigo pines and chaparral eucalyptus, dependent on fire for seed-germination, create a tinderbox environment by secreting oils or creating heaps of dry and crackly leaf-litter. [42] This is their only way to deal with the big broad-leaved light-hogging oaks that make it hard to survive down in the understorey.

DEATHWISH PINE: Um, excuse me, you are clogging up all the light.

42 F John Odling-Smee, Kevin N Laland and Marcus W Feldman, *Niche Construction – The Neglected Process in Evolution,* Princeton University Press (2013).

MIGHTY OAK: Shut up and suck on a toadstool. I'm taking in the rays, it's a hot and sunny day up here, too bad you're down in the dark. Heh! Heh!

DEATHWISH PINE: You like it hot? You like it bright? Yeah, well my friend it is gonna be very hot. Let me just grow my lower branches to cover up that No Smoking sign, sand bucket and fire hose. Heh! Heh!

The use of fire by humans can also be seen as an example of the way in which niche construction alters evolution. In *Catching Fire: How Cooking Made Us Human,* Richard Wrangham argues that cooking with fire allowed us to spend less energy digesting our food, which in turn let more blood be diverted away from the stomach and towards the brain. Thanks to the prehistoric use of fire, our stomach shrunk and our brain expanded. In the industrialised nations of the Global North, however, this process appears recently to have reversed itself.

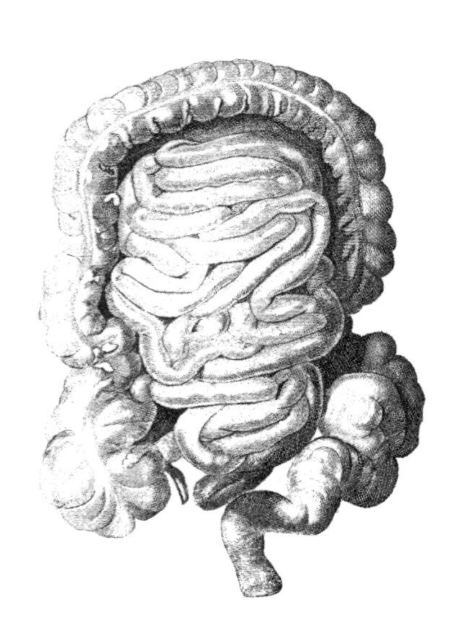

O

❧ ON THE ORIGIN OF SPECIES ❧

One of Darwin's core ideas is that human beings are the evolved product of millions of years of living in groups of social primates of one stripe or another. This inescapable evolutionary history means that we are born endowed with what he calls 'ever-enduring social instincts.' Our social instincts are biological instincts. They don't fall from some higher plane, but come from our long experience of living in social groups. They are innate, the product of hundreds of thousands of years of sharing, squabbling, falling out and making up, hugging and sulking, slapping and tickling. In Darwin's phrase, 'there is a grandeur in this view of life', but it is a grandeur utterly rejected by anti-Darwinists, most prominently Dawkins, for whom we are born selfish.

Accused of elevating selfishness into a cosmic principle or law of nature, Dawkinites claim that it is just a metaphor: 'Oh no Dawkins isn't saying humans are selfish, you've simply misunderstood. You've failed to grasp the subtle complexity of his brilliant metaphor. It's just a *metaphor,* don't you see? A *metaphor* for the apparently selfish behaviour of this particular molecule the gene. Why can't you get it through your thick head, that it is only a *metaphor*?'

One hundred English pounds to the reader who can spot the metaphor in this famous peroration from *The Selfish Gene*:

> *Be warned that if you wish, as I do, to create a society in which people cooperate unselfishly for the common good, you can expect little help from biological nature. Let us therefore try to teach altruism because we are born selfish.*

Now this is a view that derives not from *Origin of Species* but Original Sin.

❧ ORIGINAL SIN ❧

You can find versions of Original Sin in Calvinism, Anglicanism, and St Paul. At the time of the Protestant Reformation, here's how Martin Luther formulated Original Sin:

> *'[...] the infection with evil inclinations from our mother's womb! This inborn sickness and hereditary sin condemneth to eternal wrath and damnation every creature born into the world!'*

This was shortly before he lost his job at Hallmark Cards.

An idea common to thinkers such as Martin Luther, Billy Graham, Richard Dawkins, and Steven Pinker is that we are born bad, infected with nasty selfish genes and hideous animal instincts, but with effort can transcend our rotten nature. To do any good or kind action means first overcoming our base, corrupt natures. This is the complete opposite of Darwin's philosophy, as we saw just now in the Origin of Species section. For Darwin, when we do bad we very often have to transcend our sociable, empathic natures, and when we do good we go with the grain of our evolved social instincts. Self-sacrificial actions are, he says:

> *'the simple result of the greater strength of the social or maternal instincts rather than that of any other instinct or motive; for they are performed too instantaneously for reflection, or for pleasure or pain to be felt at the time; though, if prevented by any cause, distress or even misery might be felt.'*

That word misery is almost always the word Darwin uses when he imagines any thwarted instinct. It's the word he uses to describe what a goose feels if prevented from migrating by clipped wings or a cage, and to explain why the goose will throw itself at the bars of the cage until its chest is bloody. I think his connection of misery and instinct in the human context just quoted is every bit as strong. Try to deny your generous impulse, you will feel miserable because you have gone against deep biological instincts.

It's clear, therefore, that the views of a Pinker or a Dawkins do not derive from Darwin or Wallace. So where do they get their ideas then? They are detailed expression of a philosophical doctrine which runs through Adam Smith, Hegel, Hobbes, Luther, Calvin, and Saint Augustine. The Pinker/Dawkins view is perhaps most succinctly encapsulated by the ninth of the 39 Anglican articles of faith:

> *Original Sin [...] is the fault and corruption of Nature [...] whereby man is of his own nature inclined to evil, so that the flesh lusteth always contrary to the spirit.*

ÖVERKALIX

The Swedish town of Överkalix lies way up north right on the rim of the Arctic Circle. In the nineteenth century, the town was so isolated that if a harvest failed the Överkalixians starved. The parish registers show a fluctuation between failed harvest and bumper crop, famine and feast.

- 1801 famine
- 1802 feast
- 1821 famine
- 1822 feast
- 1856 famine
- 1863 feast

When the study looked at the descendants of those who'd lived through this, they discovered something incredible. The grandchildren of those who got the least food lived longest. In fact, they lived 32 years longer. The average life expectancy of those whose grandparents' slow growth period (which is 9-12 years old for boys, 8-10 for girls) coincided with a famine year, was 32 years longer than for those whose grandparents' slow growth period coincided with a feast year. This turns survival of the fittest upside down. What do we call this? Survival of the scrawniest?

Passed down in the chromosome, the grandparents' life experiences are expressed in the grandchildren's size, shape, mental and physical health.

By the way, we must never let our parents know that their life story could in any way be the decisive influence on our children's health and happiness. They cannot be allowed to take the credit for the fact that their grandchildren are less warped than we their children. After all, one of the great joys of parenthood, if not its whole purpose and meaning, comes from its being a very public damage reversal project, performed right in your own parents' face. My friend Ada told me about the following exchange with her dad, which began when her dad started on about the difference between Ada and her daughter.

'Oh, she's very well behaved, my granddaughter,' he said. 'Not like you were at that age. Always mumbling and wetting yourself. Oh no. Very confident she is. You weren't a bit like that. She's a pearl.'

'Isn't it amazing,' Ada asked her dad, 'what just a little bit of love and a modicum of praise can do?'

'What's that supposed to mean?' he replied.

'I'm not blaming you, Dad. You were bringing up kids before it was widely known that continually chipping away at a child's self-esteem would in some way damage them – be so damaging, in fact, that she wouldn't have kids herself until she was 45.'

'I told you not to hang about,' he said.

'I wasn't hanging about. My delaying parenthood was a genetic adaptation to reduce the malign influence of a father who, on school Sports Day, heckled me by singing: "Last in the sack race, you came last in the sack race."'

'I was ashamed to see the other teachers overtake you. You're the headmistress. You should be showing them who's boss.'

P

PLASTICITY

The Royal Society defines plasticity as 'the ability of an organism to modify its development in response to environmental conditions.'[43]

It used to be thought that the environment sort of lies in wait for whatever random gene mutation the organism throws its way, which it either crushes or allows. Variation proposes, selection disposes. But this neat and tidy picture has perished before a growing awareness of the extent to which surroundings mould the organism.[44]

But how can landscape create a new variety of organism? How does the dumb habitat and its critters kit out the sessile oak with thick leaves or thin, or the Arctic fox with brown fur or white?

Well, here's three examples.

#1 When a daphnia water flea detects the presence of a water bug predator called notonecta, it grows a helmet and a tailspike. Daphnia's armour comes at great cost to the rest of its development, just like a child born into violence and fear who makes an early choice to close himself or herself down. If there is no chemical trace in the water of notonecta then daphnia leaves off the kevlar, and matures in the fullness of time, into a slightly larger, if more irritatingly self-assured, adult.

43 Armin P Moczek, *The role of developmental plasticity in evolutionary innovation*, Proceedings of the Royal Society of London (2011).

44 This has led to an explosion in papers about the role of plasticity in evolution. A Forsman puts this explosion as going from less than 10 papers about plasticity in 1983 to nearly 1300 papers in 2009. A Forsman, *Rethinking phenotypic plasticity and its consequences for individuals, populations and species*, Nature (2014).

#2 A pond carries trace chemicals of a bream's scent. The nocturnal mole salamander, *Ambystoma talpoideum*, is often quite happy to stay neotenic and retain gills all its life. If however, while swimming through the pond, its keen nose detects a predatory bream's chemical trace, it will grow legs and walk out of the pond and into the undergrowth. In this scenario, the pond, by carrying trace chemicals of the bream's scent, makes the mole salamander grow legs.

#3 My third and final example of how habitat creates the observable characteristics of an organism (the phenotype) involves the spawn of spadefoot toads in ephemeral desert ponds in Tornillo Flat, Texas. Here, before metamorphosing into spadefoot toads, the tadpoles somehow calibrate from the depth and warmth of the water exactly how long they've got until the desert pond evaporates. Finding themselves in hot, shallow water – a short duration pond – they accelerate metamorphosis to become adults more quickly, at the expense of feet so small an unkind person might call them trowelfoot toads. If, on the other hand, they sense they are in deep and cool water then they take their time, developing slowly and surely into larger toads, with the signature splayed feet that any spadefoot could be proud of.

By the way, this incredible example comes from a paper written by a scientist called Robert A Newman.[45] My middle name being Alan, I have sometimes been sent invitations which were meant for him. Mostly I just binned them. But once, when the invite was to speak at the American Academy of Science conference being held at Martha's Vineyard off Cape Cod, Massachusetts, with travel and accommodation thrown in, I'm afraid I convinced myself that the invitation was actually meant for me all along, and wrote back to accept the great honour of delivering a lecture on plasticity to so august an assembly.

45 Robert A Newman, *Developmental Plasticity of Scaphiopus couchii in an unpredictable environment,* Ecology (1989).

Robert A Newman, if you're reading this, and have been wondering why your membership of the American Academy of Science has been revoked, let me explain.

Upon arriving at the Martha's Vineyard International Conference Centre, I learnt that my – our? – lecture was to be the keynote speech before a packed plenary of the annual conference of the American Academy of Sciences. I should have fled there and then, like a mole salamander hopping out of a pond full of bream. But eager to convince two thousand of the world's top scientists of my new and original theory of the role plasticity plays in evolution, hubris bested me, and the next thing I knew I was standing at the podium.

'My research into plasticity,' I began, 'suggests that if you pull a face and the wind changes you'll stay like it.'

I hadn't got much beyond this before population geneticists started throwing plastic bottles at the stage. I only managed to win the crowd back with a slew of knob gags and impressions.

I would like to publicly apologise here and now to Robert A Newman for any damage I may have done his reputation and his scientific career.

Although, that said, perhaps I'd be a teensy bit more contrite if at last year's Edinburgh Festival, a certain person who spends a lot of time in Texan desert ponds didn't happen to have done a show called *Robert Newman's Hilarious World of Spadefoot Toads in Desert Ponds*.

So, suddenly we can't seem to find that first letter of the alphabet, eh? Dropping the middle initial now, are you? Are you happy tricking comedy fans out of the price of a ticket, huh? Maybe you have spent too long studying slimy toads, mister!

There's no disagreement about the role of plasticity in development, but there is about its role in evolution. Fear of looking like Lamarck has led biologists to downplay the role plasticity plays in evolution. That fear never bothered Charles

Darwin much, and in a letter of 1881 he wrote:

'I speculated whether a species very liable to repeated and great changes of conditions might not assume a fluctuating condition ready to be adapted to either condition.'[46]

What Darwin calls 'a fluctuating condition' now goes by the fancy handle of phenotypic plasticity. The genetic constitution of an organism (its genotype) is capable of a suite of alternative set-ups depending on what conditions are like out in the world. These alternative set-ups are phenotypes. Your phenotype is how you turn out in response to environmental cues. What's clear is that nature selects not just for this or that novel trait, but for those organisms most able to vary what they do in the face of fluctuating conditions. Nature selects for flex.

Plasticity reaches down to the bone. 'But surely,' you say, 'you are not going to dispute that if you have genes for a certain shaped skull, then those genes have evolved over millions of years, and all the Sure Start early intervention in the world won't change your skull's shape.'

I'm very glad you mentioned pointy skulls, because the developmental plasticity of the wild spotted hyena's sagittal crest utterly refutes your whole argument. The pointy ridge at the top of the hyena's skull, the sagittal crest, is an attachment site for feeding muscles, as are cheekbones and forehead. In *Developmental Plasticity and Evolution,* Mary Jane West-Eberhard observes that if hyenas aren't chewing bones in the wild, then it's not only muscle that doesn't develop, but bone too:

'In animals raised in captivity, which feed rarely on bone, these feeding structures are poorly developed. Elderly captive hyenas have skulls that resemble those of [wild] cubs.'

46 This is, I'm afraid, less impressive when you read the handwritten original in the British Library's Rare Documents Room because Darwin turns out to be one of those people who puts bubbles instead of dots over his 'i's and 'j's. Plus it's on Beatrix Potter-themed writing paper, and Darwin has drawn an arrow towards a picture of the sparrows helping Peter Rabbit escape from Mr McGregror's gooseberry net with the note 'cf. mutualism between Egyptian plover and Nile crocodile??'

Or, to quote the opening line of my plasticity lecture delivered to the American Academy of Science:

'The less bones you chew the less bony your skull!'

PLOTCH PLOTCH PLOTCH

I went to an Evolutionary Psychology conference at the Royal Society to deliver a paper entitled *How Animals Evolved To Make Different Noises In Different Parts Of The Globe*. Why do Greek ducks go *papapa* and Danish ducks *raprap*? What instinct, what quirk of brain chemistry impels the Belgian turkey to go *Irka-coek-coek*, and its American cousin *gobble gobble*? Why does the Korean rooster go *Coo-Koo-Ri-Koo*, the Norwegian *Kikylicky* and the English *cock-a-doodle-doo*?

The conference organisers claimed that time constraints forced them to restrict speakers to those who weren't off their meds. So I didn't get to deliver my paper, and I was scuffing around the conference in a desultory mood, which was made still worse by the fact that every seminar and lecture I dropped in on was promulgating the dog-eat-dog version of evolution. I grew more and more dejected until I met a wonderful neuroscientist called Marie.

We started chatting and we were getting on very well, so well in fact that she began slagging off her ex-partner. What music is sweeter to the human ear than the sound of someone you fancy slagging off their ex? But the more she described her ex-boyfriend's failings, the more I couldn't stop myself from pretending to be the complete opposite.

MARIE: ...Yeah, it was just a really strange thing about him.

ROB: Oh yeah? Really strange, yeah? What was that?

MARIE: Well, it was very curious. He really hated the sea.

ROB: (*piratically*) Haar-aaar!

Getting on so well we were that she invited me to the Biologist's Ball, which was being held that very night in the Royal Society. It was a formal affair, so we had to split up, go home, scrub up, get changed. Before we went our separate ways, she said:

MARIE: Oh, and just so we don't turn up in exactly the same get-up, I should let you know that I'll be wearing a facial expression of melancholy foreboding.

ROB: Good job you told me. That's what I was going to wear. In that case, then, I'll go for brittle insouciance combined with puckish naïveté.

Thus attired, I bounded up the Royal Society steps at sunset. And oh, it was beautiful. The ballroom was hung with chandeliers, there was a 12-piece orchestra, everybody dressed to the nines, so I asked Marie,

'What's with the melancholy foreboding?'

'I've got an appointment tomorrow morning with my gynaecologist,' she replied, 'a tall, powerfully-built Finnish woman of an austere beauty. There's a strange vibe between us which manifests itself in her being somewhat rough in her handling of me.'

As Marie went into some detail describing this rather painful-sounding, invasive procedure, I crossed my legs. She no sooner saw me do this than she cried out,

'Mirror neurons! Brain scans of the anterior insula have found that the same neurons that flash when monkey peels nut also flash when monkey sees you peel nut. Same neurons that flash when someone taps your shoulder also flash when you see someone's shoulder tapped. The fact that when I describe a painful smear test you cross your legs proves that empathy is hard-wired into who we are, and that we literally feel another's pain.'

And I hadn't the heart to tell her that I was just trying to disguise an inappropriate hard-on. Instead I said:

'The discovery of these mirror neurons lends great weight to one of my core working hypotheses which is that cooperation drives evolution more than competition.'

'That is a lonely position to take, my friend.'

'Maybe so, maybe so.' But I didn't feel lonely standing shoulder to shoulder with Marie on the edge of the dance floor just as the 12-piece orchestra struck up with Shostakovich's *Jazz Suite*. I looked into Marie's eyes, and she looked into mine. And you know how sometimes, when you're the first up to dance, you can feel self-conscious, especially if you are dancing in the old-fashioned way? But I was so delighted to have met a fellow spirit that I had not a scintilla of self-consciousness, I was in ecstasy gliding around the floor. After a while I called out to Marie, 'Come and join me!' Evidently not the dancing kind, she invited me to join her instead out on the veranda, where she said:

'If you're really interested in mirror neurons, me and some of the other researchers who have worked on this project are going for a picnic at the weekend, and you'd be most welcome to join us. I'm sure they'd be only too happy to answer any questions you may have.'

'No can do, Marie,' I said. 'No can do. I'm off to Belgium this weekend to do some research in the field.'

'Oh? What sort of fieldwork?'

'Well, it so happens that there's a population of wild turkeys on the Franco-Belgian border. Now, the Belgian turkey goes 'Irka-coek-coek', the French turkey 'Glou-glou-glou'. I'm thinking... if there's been cross-breeding has this produced a hybrid language, or, like Mendelian sweetpeas, are the offspring all either discrete *Irka-coek-coeks* or *glou-glou-glous*?'

And it was good to see Marie's facial expression go from melancholy foreboding to... concern.

I took the Eurostar to Belgium, which is about as good as it gets, but we must concede, I feel, that romance has vanished utterly from train travel. I have some plans and proposals to bring it back. Top of my list is the reintroduction of the see-saw hand-operated railway trolley as used by Laurel & Hardy, Charlie Chaplin and Buster Keaton. Even if you never got to ride one yourself, just knowing it was out there plying the rails would increase the sum of human happiness, wouldn't it? To be standing on a platform at Edinburgh Waverley, Manchester Piccadilly or London King's Cross, and just to hear the tannoy announcer say:

TANNOY: The next departure from Platform 4 will be the 12.20 see-saw hand-operated railway trolley for Royston, calling at Stevenage, Hitchin, and then back again, at twice the speed followed by a large steam locomotive.

It would be beautiful!

I alight at Brussels to discover that Belgium is a fascinating country politically, riven as it is by an ethnic fissure that runs right down the heart of Belgian society. There's your Flems and your Wolloons, and ne'er the twain shall meet. If you're a Flem you are more likely to marry an El Salvadorean than a Wolloon. And there is a real concern that Belgium may not continue to be a single nation for very much longer. I heard an interview on the World Service recently with the Belgian Home Secretary and she was saying:

'Well, if Belgium is going to have any chance of remaining a single nation for very much longer then it is going to take courage, grit and determination – and these are not Belgian characteristics.'

So there I am on the Franco-Belgian border, armed with a boom microphone and a digital recorder crawling through the undergrowth towards a covey of wild turkeys. But I can't hear anything because a couple of fields away there are these hunters in waders shooting at pheasants.

I stand up and wave so they'll know where I am and not shoot in my direction. They wave back, but it's another hour before they finally break the stocks of their guns and leave.

...And that's when I became the first person ever to hear this new hybrid turkey language – neither *Irka-coek-coek*, nor *glou-glou-glou* but something intermediate. I record it and run back to the cafe, my heart thumping in excited anticipation of the front cover of science journal *Nature* that I'm definitely gonna get off the back of this. Just then the hunters come into the cafe, spot me and say:

'Hello. Hope we didn't put you off too much with all our shooting.' He raised an invisible shotgun, and mimed firing. As he did so, the next words he spoke sent a terrible shiver down my spine:

'PAM! PAM! PAM!'

At first I was at a loss to know why these onomatopoeic words cast a sickly presentiment of doom over me. All I knew was that I couldn't breathe inside the cafe. I had to get outside. I stood on the pavement gulping down cool draughts of air just as a crocodile of Belgian schoolchildren came towards me singing *Wheels On The Bus* in French.

Les rous du bus tourne et roule,
tourne et roule, tourne et roule,
Les rous du bus tourne et roule,
toute la journee.
Les essuises glace du bus font...

The wipers on the bus go...

Somehow six years' work now hung on whatever would be the next syllable to fall from their mouths.

Plotch, plotch, plotch.

Every plotch a red-hot nail in the coffin of my theory. Six years' work down the drain. And given that life is just one humiliation after another, the coping mechanism that I have devised is that even in the very instant of suffering the humiliation I flash-forward to a fantasy scenario in which everything is made all right again:

As the King of Sweden draped the Nobel prize around his neck, it was hard for Newman to believe that, but six months earlier, he'd been standing on a Belgian pavement shouting at schoolchildren the words: 'It's swish, swish, fucking swish, no wonder your country's falling apart!'

What am I gonna do now? I asked myself. Where do I go from here?

I took the Eurostar back to England to see Marie. I told her about my Belgian plotch plotch plotch debacle, and in response she gave me a gift.

'Here, I want you to have this,' she said, handing me a book. 'It's a collection of WH Auden's last ever poems.'[47]

'That's very kind of you. But why?'

'Listen to this,' she said, and read these words out loud:

As a rule it was the fittest who perished, the misfits,
Forced by failure to emigrate into unsettled niches,
Altered their structure and prospered.

'Ah yes,' I said, 'you can tell it's one of Auden's last ever poems because his poetic gifts are failing him.'

'What?'

'Unable to find a rhyme for *niches* he settles for *prospered* when there's a perfectly good rhyme going begging:

47 WH Auden, *Thank You, Fog*, Random House (1974).

The misfits, forced by failure to emigrate to unsettled niches,
altered their structure and became a new species.

Easy peasey, lemon squeezy.'

'Have you learnt NOTHING,' asked Marie, 'about tampering with the words of great poets from your Royal Shakespeare Company fiasco?'

'Oh,' I said. And then, a little later, 'Ah.'

Earlier in the year a colleague of mine was guest-curating an RSC season at Stratford Upon Avon and invited me to direct the Hamlet. Now, I've always had a problem with Shakespeare's use of mixed metaphor in the famous soliloquy and so I decided to make some changes. Your threshold tolerance for mixed metaphor may be higher than my own.

Whether tis nobler in the mind
To suffer the slings and arrows of outrageous fortune,
Or to take up arms against a sea of troubles
And, by opposing, end them?

You cannot flail the waves with a sword. You cannot shoot a pistol ball at the tide, and so I sounded out the actors as to whether they would be prepared to replace the phrase 'take up arms' with 'inflate a dinghy'. In principle they said yes they were up for it, but they didn't like the way that particular phrase scanned, they didn't like its rhythm, and so we workshopped some alternatives and came up with one which had a wonderful lilting cadence, and that is what we used on opening night. Which was also closing night, but it remains, I believe, an improvement on the First Folio:

Whether tis nobler in the mind
To suffer the slings and arrows of outrageous fortune,
Or ride a pedalo against a sea of troubles
And, by opposing, end them.

Chastened by Marie reminding me of this, I read the Auden misfit poem again. Little did I know it was soon to change my whole life, by leading to the major breakthrough in evolutionary theory known as Misfit Theory. [See **Survival of the Misfits**].

❧ POPULATION ❧

Both Charles Darwin and Alfred Russel Wallace read Thomas Malthus' *An Essay On the Principle of Population* just before they independently hit upon near identical theories of natural selection.

The Surrey parson Thomas Malthus wrote his tract to illustrate what he saw as an inescapable paradox: the better you make life for the poor, the worse you make life for the poor – and for everybody else as well. If you bring in poor relief then people who would naturally have died will live, leading to a vicious circle of resource wars and mass starvation. Why must this be? Because of the following brute fact of political economy, which there is no getting around. Human population grows geometrically, the amount of food in the world arithmetically. Result: famine, war, misery. Malthus' tract is, therefore, an excellent example of a truth identified by Bertrand Russell:

> *The worse your logic, the more interesting the consequences to which it gives rise.*[48]

The consequences couldn't be more interesting – catalysing Darwin and Wallace to theorise natural selection – and Malthus' logic couldn't be worse.

No law of nature states that the total number of humans and the earth's carrot content must diverge. The ratio of carrots to humans depends on how many carrots you plant. The planet's primary productivity is not pegged to human birthrate like the

48 Bertrand Russell, *History of Western Philosophy,* George Allan and Unwin (1947).

pound to the dollar. Famine was never yet about the carrying capacity of the planet. So far, at least, it has always been about who has a say in what is grown where and for whom. In the words of Nobel-prize winning economist Amartya Sen:

'Famines don't happen in democracies.'

Leo Tolstoy called Parson Malthus 'a monstrous mediocrity' because in the nineteenth century the only thing the Russian Empire had more of than hungry people was millions of hectares of temperate grassland composed of rich black chernozem soil, one the world's most fertile biomes, stretching all the way from Ukraine to the Far East steppe.

Op-eds claim overpopulation and global warming have put us slap bang in a Malthusian world and proved the Surrey parson's nutty sums right after all. The science is more circumspect. 'We already grow enough food,' said a recent piece in *Nature,* 'to feed the projected global population of 9 billion by 2050.'[49] Yes, I grant you, a couple of years ago, drought wiped out fully 50% of the US corn crop. But 30% of US corn is raised as fuel ethanol for use in motor cars. And 5% is grown for high-fructose corn syrup to sweeten junk food.

If climate breakdown raises the spectre of food shortages in the Global North, it is not to the bad logic but the interesting consequence we should turn; not to Malthus but to Wallace we should go for a solution.

One idea close to the heart of Alfred Russel Wallace was Land Nationalisation. We should be having a national conversation about whether or not the time has come to take all non-productive land-holdings above fifty acres into democratic control. Personally, I believe the time has come to send surveyors with theodolytes tramping over Sting and Jamiroquai's rolling Buckinghamshire acres. People say,

'Oh, but if you take away their land these people will just leave the country.'

It's FLAWLESS!

49 Brian Goulson, *De-intensify agriculture,* Nature Comment 521 (2015).

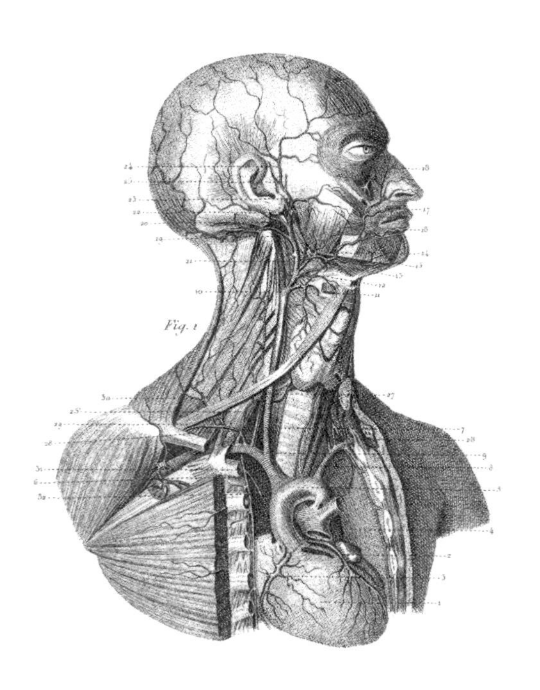
Fig. 1

Q

QUORUM SENSING

Earth has always been a planet run by and for bacteria. We are only just beginning to have the first idea about the eerie sophistication of so much of what they do, such as bacterial communication systems.

Each bioluminescent vibrio bacteria produces a chemical signal called an autoinducer. As its host, the Hawaiian bobtail squid jams ever more of these bacteria into its light organ, the vibrio reach a critical concentration that triggers the enzyme luciferase to switch on vibrio's light-producing lux genes. This is called Quorum Sensing because the bioluminescent bacteria only switch on their light when they sense they are in a quorum. This stops them wasting energy, because an individual vibrio can't produce much light, but beyond a critical threshold they glow like UV.

For vibrio, emitting light is something they only do in company. For them, as for us, there are some things you only ever do when you are with a bunch of mates.

...Although, that said, I confess it came as some surprise to learn from his diaries that when the solitary Wordsworth, finding himself alone on Westminster Bridge, composed the immortal lines:

The river glideth at his own sweet will:
Dear God! the very houses seem asleep;
And all that mighty heart is lying still!

He was actually up the top of a lamppost wearing a traffic cone on his head.

R

❧ ROMANTICS ❧

Romanticism saturates current evolutionary thinking. Early humans are only ever painted, drawn, CGI'd or facially reconstructed in a very limited set of stereotyped poses. All of them Romantic.

Look at that caveman frown like Beethoven as he strikes fire from flint. There he goes, striding across a hostile landscape of immense sweep, a cloak of hide around his shoulders, destiny in his eyes.

You will never go into a natural history museum and see prehistoric humans (or rather their depictions) do any of the following:

- pull funny faces and make silly noises to amuse a toddler.
- kvetch about back-ache.
- make up a tune while walking.
- try to talk while eating and have food go down the wrong way.
- laugh until the water she has just drunk comes out her nose.
- be in a big group of friends who find a great ledge for diving, and spend hours and hours doing somersaults and belly-flops.

It is impossible that our ancestors did not do all these things. They are *our* ancestors after all. We did not, despite what Francis Crick believed, come from outer space in the rocket ships. But the lesson from museums, documentaries, and textbooks seems to be that our ancestors were too busy evolving into us to muck about. If they'd been larking about at the swimming hole, now where would we be, huh? We'd have perished in the snow. There was no time to muck about because it was so hard to subsist. Yes, but if they hadn't been mucking about then how could they ever have become us?

The Romantic insistence on the craggy, the dour and the windswept, gets its impetus from being a reaction against another Romantic vision: the Pleistocene pastoral. All those op-ed pieces saying that we suffer from having a Stone Age brain in a high-tech world imagine an idyllic Pleistocene. Our evolutionary history, they tell us, hasn't equipped us for the fast rate of change in modern life. It used to be that a hand axe design lasted hundreds of thousands of years, but nowadays a new version of iTunes wants permission to download every half hour. And that's why we go crazy, smoke crack, and run through the streets with a meat cleaver. It's the stress of it all. Everything changing so fast and that. We can't take it. Can't keep up. Not with our poor Stone Age brains that evolved during halcyon millennia on the East African savannah, our ancestral homeland, those softly waving grasslands where time stood still.

In fact, the rate of change in the Pleistocene Rift Valley was fast and furious, stressful and bewildering. You are throwing rocks at prowling megafauna you've never seen before. You are continually on the move. There are volcanic storms and lava flows. A land bridge appears, a valley disappears. Lake floods land. Lake evaporates.[50] Vegetation zones move southward, taking the familiar animals with them. The hunter gatherer party that went off last month to look for nuts and fennel are cut off by ice, never to be seen again, until 80,000 years later, when your descendants meet at a party in Brooklyn and fall into an unexpectedly bitter argument about food miles.

A classic Romantic death was to die of consumption at age

50 Mark A Maslin et al. *East African climate pulses and early human evolution,* Quaternary Science Reviews (2014).

25. In the Pleistocene epoch, this wouldn't have enjoyed quite the same caché because at 25 you'd be the oldest member of the tribe. All the mourners would be standing round at your funeral saying:

'Well, she had a good innings. Better she went now than be a burden. There's no quality of life after 25 is there? I'd have hated to see her linger on to 27!'

❧ RUGGED INDIVIDUALISM ❧

In 1893, at the Chicago Expo, Frederick Jackson Turner presented a paper called *The Significance of the Frontier in American History*, while Buffalo Bill's *Wild West* show retold the Oregon Trail as a fight to the death between superior European settlers and inferior Native Americans, who must give way to the inexorable progress of Manifest Destiny.

In fact, as Buffalo Bill well knew, Cheyenne altruism saved the white colonists when they were trekking West and ran out of food and water and were drowning in rivers (drowning was the biggest cause of death on the Oregon Trail). Likewise, the Pilgrim Fathers were starving, their European import crops perishing in the New World soil, until the Powhatan Native Americans taught them a little soil science and handed out food parcels to tide the colonists over the winter. All this has since been written out of the story in favour of the myth of rugged individualism.

Between Frederick Jackson Turner and Buffalo Bill, the Chicago Expo spawned the cult of rugged individualism, a perfect fit with survival of the fittest, both of which became weapons in the struggle against Big Government in the United States.

Rockefeller and Carnegie hosted a banquet at New York's Delmonico's restaurant in honour of Herbert 'survival of the fittest' Spencer, who was toasted by Carnegie as the greatest philosopher of all time. As well he might for Spencer's survival of the fittest had just made laissez-faire capitalism look like the highest, noblest social justice.

The cult of rugged individualism reigned for forty years, before biting the dust of the Great Depression's dustbowl to be replaced by Franklin Delano Roosevelt's Works Progress Administration in the United States.

For a generation after the war, *survival of the fittest* was beneath contempt. Dead heroes leave no offspring – so how could you say the fittest survive when all the best and brightest had just been killed? Post-war Britain achieved the welfare state, democratic control of nationalised industries and perhaps the greatest social mobility in human history. Then the 70s brought The Great Rollback of the Post-War Settlement. Reaganomics in the USA and Thatcherism in the UK held that laissez-faire capitalism was more likely to succeed because it was closer to how things really worked in life, closer to the nature of things. Ideological outriders for these rollback projects were EO Wilson's *Sociobiology* in the United States (1975) and *The Selfish Gene* in the UK (1974).

The paradox of selfishness as a force for good in the world has a history. It comes from the Enlightenment reaction against attempts at state-intervention in social life. The Scottish economist and moral philosopher Adam Smith argued that top-down reforms don't work, whereas small selfish acts of 'enlightened self-interest' do. It's the same message you get in Bernard de Mandeville's *The Fable of the Bees* (1714):

> *Thus every part was full of Vice*
> *Yet the whole mass a Paradise.*

So long as everybody looks out for No. 1 then we'll be all right. The disasters will only come from collective effort... Except that's not how a hive of honeybees works, is it? Nobody today would describe a hive of bees as a den of vice. When bees perform a waggle dance to the other bees, it is only accidentally bootylicious. If a bee twerks, blame the flight path to the nearest nectar, a flight path whose twists and turns she is faithfully replicating. In fact, don't blame the flight path, blame yourself, you pervert. It's a bee. What's wrong with you?

'Practical men,' wrote John Maynard Keynes, 'who believe themselves to be quite exempt from any intellectual influence, are usually the slaves of some defunct economist.' In the same way, science writers who believe themselves to be dispassionately reporting what the science says are often the slaves of some defunct political philosopher. And sometimes they form a positive feedback loop, mutually reinforcing each other.

Selfish gene theory has never had less scientific respectability and at the very same time it has never had more influence. In large part because it is such a magnificent vector for free market economics. You can hear its influence, for example, in the way the business press huzzahs German intransigence over fiscal austerity for the rest of Europe, or in Angela Merkel herself, when she says:

'It doesn't actually help the Greeks if you spend your time helping them. No, you have to plunge them into the crucible of unregulated competition and from the smoking wreck of Athens, leaner, meaner, fitter – the ones that deserve to live – shall crawl.'

Marvelous Nietzschean rhetoric! Stirring stuff! Uplifting and enlivening to read! Problem is, it's not exactly how Germany became a post-war economic and industrial powerhouse, is it?

In 1946, Germany had almost all its debt written off so that it could reindustrialise, which it did with such gusto that by 1952 West Germany was $65 billion dollars in debt. Cue the 1953 London Debt Conference, at which all the nations of Europe, including Greece, including Ireland, agreed to forgive West Germany all its debt, right down to the last pfennig. Now this wasn't pure altruism of course. The European nations shared a memory of the fact that they had once before attempted to impose a policy of stringent fiscal austerity on the Germans in the 1930s. And I'm not sure if the phrase 'with mixed results' is entirely adequate or in any way fit for purpose to describe what then ensued. But there was a more communitarian ideal in international affairs at that time. As the London Debt Conference's Declaration put it, they were forgiving German debt because:

'Desiring to remove an obstacle to normal economic relations [and] as a contribution to the development of a prosperous community of nations.'

In the 1950s the Germans were not plunged into the crucible of structural adjustment that the Greeks are being plunged into now. But just as the idea that living creatures evolve to do things 'for the good of the species' or 'the good of the group' is totally and utterly the wrong message about natural selection, so what the historical record has to say about the 1953 London Debt Conference is totally and utterly the wrong message about the economy.

'Capitalism,' said Keynes, 'is the extraordinary idea that the nastiest of men for the nastiest of motives will somehow work for the benefit of all.'

This extraordinary idea enters the bones like freezing sea mist. The public sector is a sickly dependent, only perfused and kept alive by dynamic, go-getting private power. Look at Google, poster child of free market innovation. There's an example of what happens when private enterprise is allowed to slip the leash of unwieldy Big Government. Research and development of the algorithm behind Google's search engine, however, was funded by a National Endowment grant, or in other words, by the stultifying dead hand of democratic finance and the public sector. The only private thing about Google is its profits.

But (to borrow a phrase from Shakespeare scholar James Shaprio) let's not let truth get in the way of truthiness.

A
B
D
C
G
S
E
S
H
P
V
B
C
E
H
D
G

S

SEX

In *On Human Nature,* sociobiologist EO Wilson argues that while sex equality legislation may be laudable, such laws go against the grain of human nature. Laid down over millions of years of natural selection, our gendered psychologies are simply not amenable to PC diktat. Sorry, ladies, but that's just the way things are:

> *Even with identical education for men and women and equal access to all professions, men are likely to maintain disproportionate representation in political life, business and science. Many would fail to participate fully in the equally important, formative aspects of child rearing.*[51]

What woman would be allowed to get away with logic like this? These two sentences simply cancel each other out. If men are failing to do their share of childcare, then how can there be 'equal access to all professions' for women? Who are these sex-neutral aliens from the planet Neptune bringing up the kids? Has Wilson had a positive sighting of Francis Crick's space aliens? Men have been on an undeclared childcare and housework strike for years, and this has damaged women's chances. (Imagine the ironing that EO Wilson could have usefully done instead of writing *On Human Nature.*)

Notice how EO Wilson is unable to follow the logical entailments of his own thought-experiment. If the starting premise of Wilson's scenario is an imaginary society in which women have 'equal access to all professions' then the only way

51 EO Wilson, *On Human Nature,* Harvard University Press (1978).

men could possibly enjoy 'disproportionate representation in politics, business and science' would be through breaking the law by leaving their toddlers unattended. Wilson's fantasy society would presumably have a legal system, courts and jails, and if 'failure to participate fully in the formative aspects of child-rearing' were a civil offence then it would be impossible for men to continue to exert a disproportionate dominance of all the top jobs.

Of course, if there were free creches for all, then these men would not have to go to jail. But free creches for all could only be paid for by scrapping male-dominated asset-stripping jobs in the financial sector. After all, you cannot afford to have the City of London *and* social justice.

Wilson describes political life, business, and science as if they were fixed things in the world like the nitrogen cycle. But again, if the starting point of his thought-experiment is an equal world then political life, business and science would be utterly transformed, so much so as to beg the questions:

Whose political life?
Which businesses?
What type of science?

When it comes to taking care of business, EO Wilson seems to envisage women failing to make a go of managing their local Hogs & Hooters or Spearmint Rhino. He pictures an over-promoted woman editor explaining falling sales figures to the board of *Guns N Ammo*. If the all-female board of a nuclear weapons silo makes regrettable overtures to the peaceniks sitting outside the fence, then there's something wrong with the XX chromosomes, because there's nothing wrong with a nuclear weapons facility.

In fact, nukes make a good barium meal to test EO Wilson's proposition. Everyone from CND to *Bulletin of the Atomic Scientists* agrees that women oppose the existence of nuclear weapons more strongly than men.[52] How fares the politics, business and science of nuclear weapons when half the members of every parliament are women?

Allowing women equal access to the British government, I suggest, would change the defence policy of every political party in Britain except the Greens. (Yes, the SNP want to get rid of Trident nuclear submarines, but they also want to stay in NATO, the nuclear-armed tanks on the UN's lawn).

What about science? Science was, you recall, the third of his triple-whammy against the doomed naïveté of sexual equality. Once again, EO Wilson talks about science as if it were some kind of free standing thing, operating a neutral process of natural selection in its remorseless weeding of enfeebled XX chromosomes from the academy.

But science does not stand outside of society. Take the Royal Society, for example, that 'fellowship of the world's most eminent scientists and [...] the oldest scientific academy in continuous existence.' In the 1600s, the Royal Society was set up on these lines:

'What is feminine [...] [shall] be excluded from the Society's philosophy,' wrote founder member Henry Oldenburg in a letter to Robert Boyle, pledging that the Royal Society would 'raise a manly philosophy'. This was off the back of Bacon talking about how the new science promised to usher in: 'a truly masculine birth of time.' For centuries the Royal Society has held true to Oldenburg's vision. Not until 1945 did it admit its first woman. And in 2014 just one in twenty Royal Society research fellowships went to women.

But for EO Wilson, even if the Royal Society had been set up as a ladies science college, with a remit that what is masculine should be excluded from the Society's philosophy, men would *still* be winning nineteen out of twenty research fellowships. Male and female inequality is in our genes, he says. Stop trying

52 Rosemary Chalk, *Women and the National Security Debate*, Bulletin of the Atomic Scientists (1982).

to explain it away with wishful thinking. It's who we are. It's what we do. And why is it what we do? Because it is what we have always done. Right from when we first came down from the trees. Except the latest science suggests otherwise. Far from being an invention of the last few decades, sexual equality now looks to have been the rule in prehistory, at least until the advent of sedentary agriculture.

In May 2015, the journal *Science* published a groundbreaking paper, arguing that 'increased sex egalitarianism in human evolutionary history may have had a transformative effect on human social organisation.'[53]

The paper was written by a team of anthropologists who studied two contemporary hunter-gatherer tribes, the Agta in the Philippines and the Mbendjele Yaka in the Congo, and found the social organisation of each tallied with their computer model of what happens when the sexes have an equal say in where they live. When men make the decision on their own, you have tight hubs of closely related sibs, but when men and women have an equal say in where to live you find settlement spread over a wider geographical area – 'I'm not living near your family. I don't want your mum sticking her nose in'. This leads to a broader social and genetic array, which has knock-on effects:

'[S]exual equality may have proved an evolutionary advantage for early human societies,' writes ace *Guardian* science correspondent Hannah Devlin in an interview with the paper's authors, 'as it would have fostered wide-ranging social networks and closer cooperation between unrelated individuals.'[54]

This may also have led to the sharing and spread of technological innovations. But, of course, that's man stuff.

53 Mark Dyble, Andrea Magliano et al. *Sex equality can explain the unique social structure of hunter-gatherer bands,* Science vol. 348 (2015).

54 Hannah Devlin, *Early men and women were equal, say scientists,* Guardian (May 14th 2015).

SPECIATION

One of the emerging themes in evolutionary biology is that the environment plays a far larger role in natural selection than was previously thought.

The direct action of the environment's central role in speciation can be seen in how something very large – the formation of the Grand Canyon – led to something very small – a new species of antelope squirrel.

While not as old as the hills, antelope squirrels are older than the Grand Canyon, which formed three million years ago. Three million years and one day ago, a single species of antelope squirrel gamboled on opposite sides of a finger-width fissure in the ground. When fault yawned into canyon, two halves of the antelope squirrel population found themselves looking at each other across a chasm. Perhaps only poetry could heal the rift in their hearts.

What once we had, now is lost,
Sundered by forces primeval.
The chasm between us will never be crossed
Not even by Evil Keneaval.

Two distinct species of antelope squirrel, white-tailed and Harris, live on either side of the Grand Canyon. The speciation had nothing to do with survival of the fittest, nothing to do with any competitive advantage in having a white tail or in being called Harris. All it took to make a new species was the sudden opening up of a great big hole in the ground.

SQUID-VIBRIO SYMBIOSIS

The Hawaiian Bobtail Squid lives in the coastal waters not just of Hawaii, but throughout the Southern Indo-Pacific as far down as Australia. Every evening, the nocturnal bobtail squid hoovers bioluminescent bacteria called vibrio into its light organ. All night long, the brightly glowing vibrio act as the squid's camouflage. How can bright lights be camouflage? Sharks and other predators are looking for the shadow cast by the squid. Light sensors on the squid's mantle measure how much moonlight in coming down, the bobtail squid projects from its light organ the same intensity of light, perfectly matching the light below to the light above. What the shark sees looks like a single uninterrupted moonbeam. It's like an Invisibility Cloak.

If the moon goes behind a cloud the bobtail squid dims its lights. If there's a storm with thunder and lightning, the bobtail squid can actually flash its lights to mimic the lightning. It's an incredibly responsive organism. Although that said it was a very long night for any bobtail squid unlucky enough to be in Sydney Harbour on the night that Jean-Michel Jarre played an open-air gig.

SQUID: What's that noise? Is it a Beluga whale? Those flashing lights? Old-timer, can you tell us?

OLD-TIMER: Last time I heard that sound the entire shoal was wiped out. All except me.

SQUID: It is a typhoon? An electrical strom?

OLD-TIMER: Worse. It's ambient-electro concept synth with a live laser display.

SQUID: Can you remember the lighting cues?

OLD-TIMER: All I remember is when it gets to the diddlydeedee bit, that's when the laser harp comes out.

SQUID: Laser harp?

OLD-TIMER: Yeah, that's when he shoots revolving neon beams of red, gold and green into the night sky. That laser harp blew our cover, the sharks zoomed in and I found myself swimming through the blood of the shoal. Oh look! There's a pod of bottlenose dolphins to starboard. We're done for! It's happening again!

SQUID: Steady as she goes, Old-timer, I've got an idea. If we all swim in a corkscrew dive we'll mimic the revolving beams. Here comes the diddlydeedee bit. Stand by for corkscrew dive. On my command... Now! Corkscrew dive! Corkscrew dive! Outstanding! Here it comes again. Eyes to me, and... Cobra roll! Demonic knot! Wraparound corkscrew! Butterfly inversion! Inline twist! Sea serpent loop! Double dip and double up!

OLD-TIMER: The dolphins are swimming away!!! Squid-vibrio symbiosis give techno synth licks!

SURVIVAL OF THE MISFITS

New species originate from tiny, isolated, unstable populations of struggling misfits at the ecological fringes (Mayr, 1964). The gene pools of tiny isolated populations are inherently unstable (Tattersall, 2003). When the going gets unstable, the unstable get going (Billy Ocean, 1986). One day, geospheric churn fells the fittest and among the misfits' suite of mutations is the must-have adaptation that propels them centre-stage. But it is only through their long apprenticeship of being least fit in the old habitat that they become fittest in the new. Therefore instead of survival of the fittest, it is more accurate to describe speciation as survival of the misfits (Newman, 2015).

Misfits generate new species. The fittest never do. Well-fitted populations have stodgy gene pools in which any and all mutations are quickly annulled. Trouble is, you're only as fit as your last habitat. The fittest fit too well, that's their problem. Pigs in clover they may be, but what happens when that clover starts to move uphill or gets too chewy to eat? Change the scene and they're gone. Stable, well-fitted, unwieldy, large populations can't respond quickly to sudden ecological change. Tiny, unstable populations can.

If all successful species today were once tiny populations of struggling misfits, might this explain the lack of intermediate species in the fossil record? If so Misfit Theory plugs the biggest gap in Darwinism.

When Misfit Theory is discussed at international science conferences, however, speaker after speaker denounces the theory on the ground that misfits are just a sub-set of specialists, and specialists lose out to generalists in times of shock, such as a sudden ecological cataclysm.

Generalists are any population of what you might call spread-betters. Not so exquisitely adapted to any one particular niche, the generalist is for that very reason better able to adapt to a new and unpredictable. A slow loris, say, while not particularly expert at peeling lychees, climbing trees or prising open a nut – can make a pretty good fist of all three when she has to.

Specialists, by contrast, are exquisitely adapted to that particular habitat the flood just swept away. A specialist might be a hammer orchid evolved to be pollinated exclusively by those thynnid wasps that are now being shovelled out of storm drains and swimming pool pump filters fifty miles away.

In defence of Misfit Theory, I'd say that misfits are topical specialists. By sheer fluke, yesterday's liability becomes today's must-have adaptation. Consider the woolly mammoth. They were already woolly long before the temperature plunged, back when wooliness was a mutation they could have done without – especially in the summer months. The Ice Age selected the woolly mammoths, while the cotton-lycra mammoth perished in the snow.

Can we see Misfit Theory at work as a cultural force in human society? I believe we can. In the last decade the major environmental victories, for example, have not come from political mainstream, but from marginalised, misfit groups on the edges like Reclaim The Power.

They came to most people's attention, I guess, in 2009 with their seminal action Climate Camp In The City, which was a precursor of Occupy Wall Street. Climate Camp in the City occupied Bishopsgate, where they set up pop-up tents, compost loos, a farmers' market and raised beds growing brassica and runner beans, and it was brilliant because there you are, right in the heart of the City of London, and suddenly you've got all these drug-fuelled anarchists bent on social destruction looking out their office windows… at a future with a future.

From there, Reclaim The Power went on to do actions against Kingsnorth and West Burton coal-fired power-stations, followed by an attempt to shut down Ratcliffe-on-Soar power station (which led to the unmasking of a vast network of police spies involved in sexual exploitation). These actions produced a straight-up political victory: no new coal-fired power stations in the UK.

As George Monbiot says, 'Social change comes not from the centre but from the margin.'

Is there is an analogy here with the proliferation of genetic novelty in marginal populations rather than among dominant species?

'Very small populations are much more genetically unstable than large ones,' writes Ian Tattersall, 'and they provide optimal conditions for incorporating the randomly arising genetic novelties that furnish the very basis of evolutionary change.'[55]

Hanging on to this insight certainly makes me feel less despondent when I'm one of only three people who've turned up at a community centre on a wet Tuesday for a direct action

55 Ian Tattersall, *The Strange Case of the Rickety Cossack,* Palgrave (2015).

training workshop. On such occasions, I look around the room and say:

'Yes, we've got the system right where we want it!'

SYMBIOGENESIS

Symbiogenesis is the idea that new species originate through symbiosis. To put it another way, new species come about when different species come together.

In *The Origin of the Nucleus* (1905) Russian botanist Konstantin Mereschkowski speculated that the first cell with a nucleus was formed when 'micrococci invaded a bacterium [and] lived as symbionts.'

Mereschkowski's views fell out of favour for most of the twentieth century, but in the twenty-first century, *Scientific American* described his hypothesis as 'tantalizingly close' to recent discoveries of how the first multicellular organisms evolved from free-floating archaea and bacteria.

In *The Symbiotic Planet: A New Look At Evolution,* Lynn Margulis argues that more species originate from symbiogenesis than from random mutation:

'No species existed before bacteria merged to form larger cells including ancestors of plants and animals... Symbiogenesis brings together unlike individuals to make large, more complex entities [...] Life did not take over the globe by combat but by networking.'

Organelles, the subunits of a cell, may once have been free-floating organisms of their own, which swallowed one another. Among our organelles, mitochondria have their own DNA and RNA and still reproduce in their own fashion – just in case their human host turns out to be a mere flash in the pan.

Margulis showed symbiogenesis at work in everything from translucent seaweed mats on the shores of Brittany, France to mixotricha paradoxa in the guts of termites in Darwin, Australia. But it took the wreck of a few submarines to clinch Margulis and Mereschkowski's symbiogenesis hypothesis.

In 2014, shards of yellow submarine were found floating on the South Pacific, north-east of New Zealand. The shattered debris was the wreckage of deep-sea submersible Nereus, which imploded 6 miles down in the Kermadec Trench, its rivets popped by 16,000 pounds per square inch of pressure. (Walking around on earth we are under 15 pounds per square inch of pressure – unless you've just been chucked when it doubles by the hour.)

Before buckling under pressure, ROVs (remotely-operated vehicles) like the Nereus take cross-sections of deep-sea hydrothermal vents. These cross-sections reveal distinct layers of bacterial and archaeal life in separate temperature zones up and down the inside of each smoker spire.[56] From these cross-sections has come a story of the origin of species, close to Mereschkowski's vision. And it goes like this.

Life as we know it started when a deep-sea chimney snapped in half about two billion years ago. Smoker spires are too hot for seawater organisms to survive, and the ocean too chilly for chimney-dwelling hyperthemophiliacs. But snap the chimney and seawater flows into the smoker spire, hyperthermophiles gush out into the ocean, the extremists find a common mean. Symbiosis begins.

56 Kormas KA, Tivey MK, Von Damm K, Teske A, *Bacterial and archaeal phylotypes associated with distinct mineralogical layers of a white smoker spire from a deep-sea hydrothermal vent site (9 degrees N, East Pacific Rise)*, Environ Microbiol (2006).

If that smoker spire never broke then nanoarchea equitans (which likes 70°C seawater) might not have met the bacterium ignicoccus hospitalis (which likes 90°C in the spire) and then where would we be? For their symbiotic dance may have led to the first ever nucleated cell.

Nanoarchea equitans is the smallest organism ever to have its genome sequenced. This sequencing revealed its H3 protein to be a dead ringer for the H3 in histone. Histone is the protein spindle in the nucleus of our cells that tightly spools up two metres of DNA. To find H3 floating free at a hydrothermal vent supports the notion that each human cell is a jar of prehistoric pond-dipping.

I like this story but I wonder whether the idea that everything starts with smoke stacks and furnaces only seems natural because of the Industrial Revolution.

What could be more fitting for a Briton than a vision of life that starts with a row of chimneys, just like *Coronation Street?*

T

❧ TENNYSON ❧

In *Love Song* (from *Part Troll*), Bill Bailey sings from the point of view of a bitter, rejected lover, who looks at nature and sees only how *'the deer, now blind, stumbles into a ravine.'* Bill Bailey's masterpiece parodies the pathetic fallacy, when we project human emotions onto nature: sombre mountains, happily dancing flowers, trees holding out welcoming arms for returning owls. The most full blown pathetic fallacy in all literature comes from Tennyson's 1849 poem *In Memoriam*, where a capital N Nature is personified as a mad bitch in frenzy, a cruel, destructive goddess incensed by the very existence of human love:

> *Man, her last work*
> *[...] trusted God was love indeed [...]*
> *Tho' Nature, red in tooth and claw*
> *With ravine shriek'd against the creed.*

Unlike those who 'reject the implications of Tennyson's famous phrase,' declares Dawkins, 'I think "nature red in tooth and claw" sums up our modern understanding of natural selection admirably.'[57]

Deal with this truth, ye who dare! For Dawkins, Tennyson is being deep and unflinching, rather than just pessimistic and mopey. (I'll never understand why being pessimistic is seen by so many people as being somehow deeper than being optimistic. The shallowest people I know are always obsessed with dark stuff.) But are people bravely divesting themselves of

57 Richard Dawkins, *The Selfish Gene - Why Are People?*

illusion when they see nature as red in tooth and claw? Or are Tennyson's fantasies as melodramatic as King Lear's?

When one of Lear's daughters refuses to give his drunken posse houseroom, the old king rages:

It will come:
Humanity must perforce prey on itself
Like monsters of the deep.

Now, a stuck-up daughter not letting her dad park all his friends in her new house does not usher in a world of human cannibalism. Nor does Tennyson's friend Arthur dying of a cerebral haemorrhage in Vienna make nature red in tooth and claw. The biosphere goes on being more green in stalk and leaf than not. Blue-green cyanobacteria are untroubled by Arthur's death and carry on their lateral gene transfers as before.

What Tennyson's image actually sums up, far from being 'our modern understanding of natural selection', is a Creationist idea of evolution, which he got out of Robert Chambers' *Vestiges of the Natural History of Creation* (1844). Earth is the shadow of the valley of death, nature a gross insult on transcendent human virtues of love and spirituality, but fear not for after the resurrection we go to a place where the lion lies down with the lamb and nothing is red in tooth and claw anymore.

Our modern understanding of natural selection has to deal with the less Romantic fact that the dry weight percentage of biota to which the description red in tooth and claw might reasonably be said to apply is at the level of statistical error.

On how many of the phylogenetic tree's welcoming arms will we find nature red in tooth and claw? The very trunk? The big branches? The over arching canopy? Let's do some tree-climbing and find out.

On the phylogenetic tree, if you climb past the Bacteria branch, shin your way along the Archaea branch, and go way out on the Eukarya limb, up past the Trichomonads, flick the Flagellates from your face, haul yourself up through the Ciliates, and carefully part the thick green leaves of Plantea, there at last, nestled between Myxomycota and Fungi you will find the tiny twig called Animalia. Here you will find amniotes and arthropods with claws but not much in the way of teeth.

And where animals do have teeth, like lorises or gorillas, how much time is spent eating fruit and leaves and how much is spent drawing blood? Wolves have teeth and claws and like their meat rare, but the problem is they also possess – in Mary Midgley's phrase – all the domestic virtues. Not that there were any wolves on the Isle of Wight, where Tennyson lived. And you have to wonder whether the British devotion to the pathetic fallacy of nature being red in tooth and claw might perhaps be linked with our lack of any wildlife more dangerous than the adder.

Among Animalia, by the way, bee species outnumber mammal and bird species combined, so if Tennyson really wanted to eulogise the death of his friend whilst also being true to the living world, then he should have written of nature yellow, black and stripy and hanging around the compost bin.[58]

58 *Bee Species Outnumber Mammals and Birds Combined,* American Museum of Natural Science (2008).

U

❧ UBUNTU ❧

The neocortex is the extra layer of grey matter wrapped like tree bark around the cerebral cortex. The social brain theory came about when researchers discovered that among primates the bigger and more complex the group the bigger and more complex the neocortex.

If you're a chimp in a group several hundred strong, then you need to keep track of who's related to who and how closely, so that you can work out who you need to groom around here to stop the big feller from snatching your mum's food all the time.

Did the human neocortex evolve in the same way?

Was our sprawling and complex neocortex produced by living in sprawling and complex societies? If so then human evolution can be admirably described by the Xhosa concept of *ubuntu* - which translates as 'I am, because we are.'

❧ UG ❧

Commonly thought to be the first word ever spoken by human beings. I recently presented a paper to the *Anthropological Review* arguing that this may have been a primitive attempt to say the word 'hug'. I await a response.

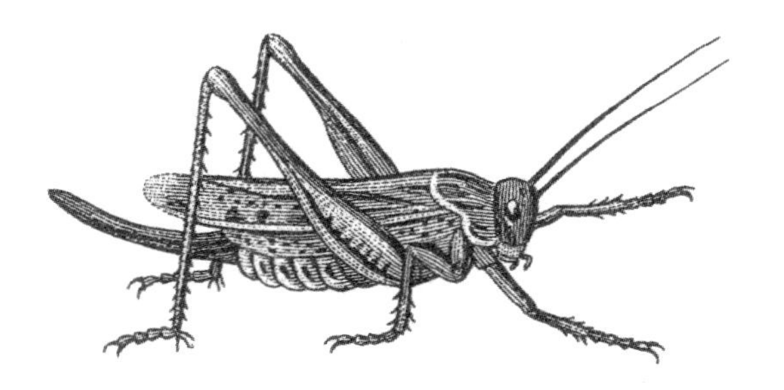

V

VARIATION

'Variability,' wrote Darwin in *On the Origin of Species,* arises 'from the indirect and direct action of the external conditions of life, and from use and disuse.' Just as his long shot on our African origins had to wait a century for confirmation, so Darwin's long-range punt on what causes variation is only now starting to look very near the mark. Until quite recently it was believed that variation came solely from copying errors during cell division. Copying errors were never very convincing as a sufficient cause for natural selection. Awareness of this vulnerability led neo-Darwinists to aggressively patrol this weak point. Doubts about the dogma of random genetic mutation were met not with defence but attack. Skeptics were accused of being too lily-livered to endure a random universe. Only the proud neo-Darwinist, he alone, was able to breathe the thin air of negative capability, to look around at a random universe and call it home.

Behind the aggressive patrols lurked a guilty knowledge of how abruptly they had split from Darwin. This guilt the neo-Darwinists dealt with by deploying what EP Thompson once called (in another context) 'the massive condescension of posterity.'

Poor old Charles Darwin, they'd lament, if only he had known about Mendelian genetics then he wouldn't have stumbled about in the dark banging his head on antiquated beams, thinking that development and environment are somehow factors in evolution. If only the dear, sweet, kind but tragically misguided bald old bastard had lived to see August Weissmann prove discrete genes to be immortal.

The neo-Darwinist pictures himself floating in a spacesuit behind the Down House bookshelves, like Matthew McConaughey in *Interstellar,* shouting 'DNA,' as poor benighted Darwin bimbles about with specimens of barnacles.

'The idea that all DNA changes arise through random mistakes is wrong,' claim Eva Jablonka and Marion Lamb. Environmental stress has been shown to give rise to mutations. Mutation is often, therefore, far from random.[59] The recognition that mutation is a response to external events is part of the intrusion of the outside world into the playpen of Genes R Us evolutionary theory.

But even though **epigenetics** has now shown how heritable traits can be transmitted without altering DNA sequence at all, I for one refuse to be as condescending to neo-Darwinists as they've been to everybody else. After all, their elementary conceptual blunder was based on lack of evidence. Some things were not known way back when they formed their primitive beliefs about discrete germline cells. The poor fools simply didn't know how methyltransferase enzymes at CpG sites silence genes. But they were so sweeeeet with their adorable conferences on Random Genetic Mutation, and all their liddle-widdle powerpoint graphs with those cutesie-wootsie lines, emanating from gene to every single living thing in the whole world ever, like the rays of the sun, while innocent the whole time of how S-adenosyl-l-methionine's electrophilic attack activates the C(5) atom. Bless!

59 Eva Jablonka and Marion Lamb, *Evolution in Four Dimensions,* MIT Press (2014).

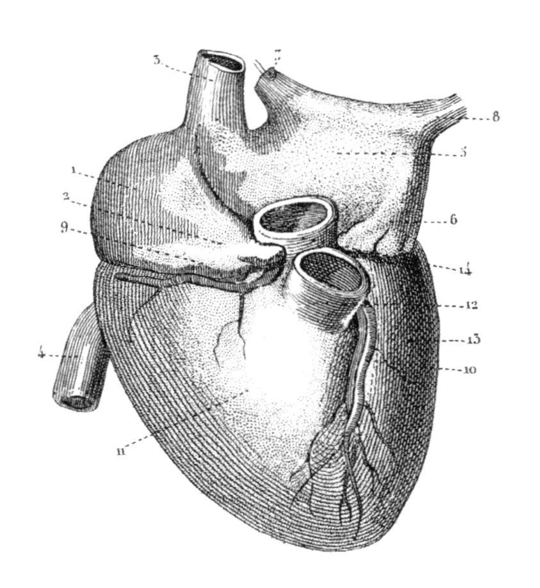
3
7
8
5
1
2
6
9
14
12
13
4
10
11

❧ WILDWAY ❧

As the winters grow soggy and the summers scald, bears and wolves are going to want to come north, the reindeer and beaver to trek south. If we build a Wildway – an intercontinental wildlife corridor, a vast, connected landscape – then these magnificent creatures will grace us with their presence.

Examining some sea-charts over the weekend, however, it was forcefully borne in upon me that Britain is completely surrounded by water. Indeed, it is not over-stating the case, I fear, to say that it is an island. The Wildway will, therefore, need a Botanic North Sea Bridge from Peterhead to Stavanger, and/or a Botanic Channel Bridge from Dungeness to Calais.

Ah, but where will the money come from? Which multi-billion-pound state subsidy to a transnational oil company do you plan on cutting to pay for your fancy Wildway? Which gas provider will be going without their welfare payments this year?

In a 2013 Finance Bill, the government gave a three billion pound slice of the welfare pie to those inveterate benefit scroungers oil and gas transnationals, with which to sink speculative boreholes in the West Shetland Deepwater Basin. On top of the three-billion handout there's another half a billion earmarked for what's called Derrick Dismantlement – who I thought played bass for UK Subs, but which turns out to be the retrieval and scrappage of tapped out drilling rigs and indemnifying these corporations from any clean up costs in the event of a toxic spill.

❧ WOLF ❧

To initiate play, wolves do what is called the canid play bow. Wolves that exploit the canid play bow, by using it as a ploy to attack another wolf, are ejected from the pack. So too are selfish wolves who don't share food equally. Philosopher Simon Blackburn points out that exile is a death sentence for a wolf because they are pack hunters.

And here's a question: how exactly do you pass on your selfish genes if you are banished from the pack?

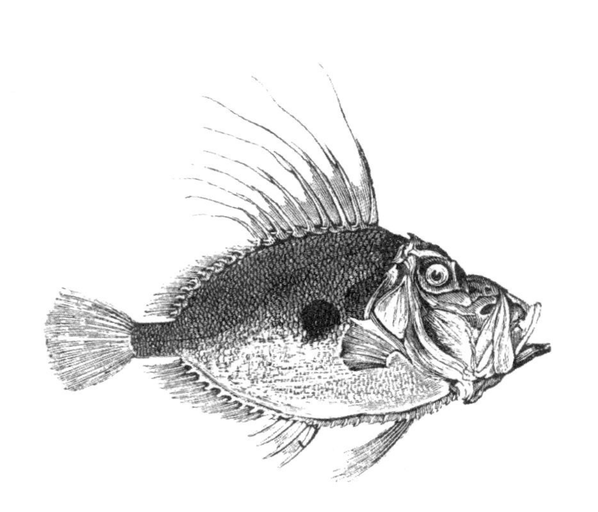

X

XENOPHOBIA

Jared Diamond is a Pulitzer-Prize-winning author, social historian, geographer, ornithologist, evolutionary theorist and anthropologist who has spent years among the indigenous people of Papua New Guinea. His *Guns, Germs and Steel* forever altered my understanding of history, of how what happened happened the way it did and not some other way. And so I regret having to disagree with an argument he makes in *The Third Chimpanzee: On the Evolution and Future of the Human Animal,* which is that humans have an innate fear and hatred of strangers. In fact, I almost can't imagine that he of all people should say this, since he has spent his whole life seeking out people who live lives very different from his own, people he has always approached with keen, frank, open curiosity. I sort of can't believe his heart is in this idea of innate xenophobia. But his book is out there, lots of people are reading it and I want to counter what he says:

> *'Our xenophobia – our fear and hatred of strangers – was manageable only so long as we lacked the means to destroy ourselves. Now that we possess nuclear weapons it may be best that we learn to see ourselves as members of a shared worldwide culture. Loss of cultural diversity may be the price we have to pay for survival.'*

Are wars really caused by xenophobia? If they are, then why are civil wars the bloodiest? By far the deadliest conflict in US history remains the Civil War. On September 17th 1862, nine times as many men were killed at the Battle of Antietam, than the United States lost during D-Day, its bloodiest day of the whole Second World War.

At Antietam, poor white Lutheran farm boys called Vernon from north-west Frederick County, Virginia fought poor white Lutheran farm boys called Vernon from south-west Frederick County, Virginia. Probably neither Frederick County Vernon could quite explain the shudder of revulsion that went down his spine when he saw the other Frederick County Vernon. But now in the twenty-first century we can explain this shudder. We now know that each feels an atavistic biological instinct. Both Vernons are driven by a deep-seated evolutionary aversion to the strangeness, the ineffable other-ness, the un-Vernon-ness of the other Vernon. It's not conscious. It can't be controlled. It runs too deep for that, as you'd expect of an evolutionarily acquired fear of those we're related to through our Aunt Emmy-Lou who fries up the best cornpone and flannel cake in the Shenandoah Valley, and who always used to get them two Vernons mixed up, 'til one time she took a notion to tie a grey ribbon to one Vernon's wrist and a blue ribbon to the other. Pretty soon, friends of each Vernon started dressing up in *whole suits* of grey or blue. Then they squared off at Cousin Ethelberta's wedding reception, punches were thrown, shots fired, all hell broke loose, and the next thing you knew, the grey Vernons wanted to secede from the Union!

Wars are not caused by innate xenophobia – which is not to say that scapegoating outsiders isn't a dismayingly effective way to gain political power, not least when those doing the scapegoating buy their ink in barrels.

Myths about the fearful strangeness of Africans, for example, helped make slavery possible in the first place (and no slavery, no Civil War).

Myths about Africans haven't gone away. They persist for example in biologistic fantasies which attempt to explain the Rwandan genocide as an eruption of innate Hutu-Tutsi rivalry, an ancient blood antipathy stretching back into the mists of prehistory. Haven't they been at each other's throats since time immemorial? Don't one tribe slaughter the other because of different tribal physiognomies – the squat, beefy Hutu driven out of their minds by the very appearance of the tall willowy Tutsi? Isn't their eternal enmity a sort of Cain versus Abel, farmer versus pastoral nomad?

In his masterful account of the Rwandan genocide, Phillip Gourevitch describes how he got so sick of hearing these whacking great clichés that he did some research and discovered that the earliest incidence in the entire historical record of any Hutu vs Tutsu violence was in 1959.[60]

'There are two types of story,' said Tolstoy. 'Man goes on a journey. Stranger comes to town.' For Jared Diamond, these stories are all in the crime/murder genre. But often as not they are love stories. For proof of which just look at how genealogists are always finding out how each of us – be we never so white or black – turn out to have what Nabokov calls 'a salad of racial genes'.

If humans have xenophobia like bats have sonar, then why the need for so many laws over thousands of years of legal history against miscegenation and marrying out? What enthusiasm were those laws trying to curb, if not our ineradicable curiosity about strangers and sexual longing for foreigners? (Reader, I married one.)

'Oversexed, overpaid and over here,' was the British moan about American GIs stationed in England in World War 2. But that was mainly the men. For the British women it was a different kind of moan. When exotic, brilliantined GIs strutted into Lakenheath Village Hall's Friday Night Hop, the local Englishwomen's guttural growling was not a territorial snarl.

I best declare an interest here, and confess that without one such American GI copping off with a local Englishwoman, I

60 Phillip Gourevitch, *We Wish To Inform You That Tomorrow We Will Be Killed With Our Families*, Picador (2000).

wouldn't be here, and you wouldn't be reading a long argument about the ways in which historical happenstance disproves biological determinism.

When Jared Diamond talks about loss of cultural diversity being the price *we* have to pay, its sounds warmly inclusive, a shared common project involving all the people of the world.

But exactly whose customs will have to go and whose will be allowed to stay? Who gets to carry on just as they are, and who has to change their ways and become part of a standardised, global monoculture? I'm pretty sure it won't be us who have to come into the reservation. We built it. But cultural diversity has to go because we now have nuclear weapons.

Wouldn't it be simpler just to get rid of the nuclear weapons? Especially when we know that a civil war in a single worldwide community would be the worst of all possible wars?

For Jared Diamond, xenophobia and Intercontinental Ballistic Missiles are fixed, immutable facts of life. Cultural diversity, on the other hand, is a moveable feast.

Sure about that?

If history shows us one thing, it's that people will die rather than give up their way of life. However irrational we may seem to you, however obviously superior you think your way to be, however much better your appliances and record-keeping, however more just your laws, we would rather live in our way, with our own laws and in the place of our own choosing. We are the immoveable object against which the irresistible force of Enlightenment universalism must always bend or break.

Y

YELLOW KELP

'*Paleontologists Unearth Earliest Known Dinosaur Stickers'* ran a front page headline in *The Onion* last year. This excavation confirms what I have long suspected: that to a greater or lesser degree we all entertain a sticker book view of evolution.

In a sticker book you have your background habitat, a swamp, say, or a mountain onto which you stick your alligator or yak. But the biosphere is no static background, not even when it appears as sempiternal as a forest of yellow kelp (*Laminaria groenlandica).*

There can't be many better fits in nature than yellow kelpfish in a forest of yellow kelp. Kelpfish wave with the current like kelp blades. They have been flitting invisibly through North Atlantic kelp forests since before sticker books were even invented. They've even got white spots in imitation of the bryozoans found on yellow kelp. All is going swimmingly until Arctic ice-cap melt lowers sea salinity, and, here and there, *Laminaria groenlandica* gives way to a type of kelp that needs less salt in its sea, the purple *Ecklonia radiata.*

It has taken millions of years for yellow kelpfish to look this much like blades of yellow kelp, but against this new purple kelp they stand out like a bon-bon on a hotel pillow, while down in reception the sea-otters are checking in.

The kelp forest gave no sign it was going to suddenly look like a Persian Ironwood in autumn – until one day it just did! Oh, to be a chromatophoric cuttlefish, mimic octopus, or pair of courting seahorses changing colour as they dance.

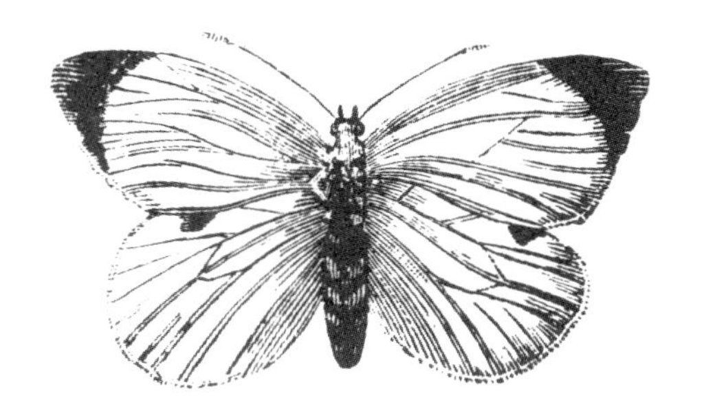

Z

ZOONOMIA

Charles Darwin's grandfather, Erasmus Darwin, was an Enlightenment genius. Around the time of the French Revolution he invented a hydrogen-powered rocket engine and theorised that humans shared a common ancestor with orangutans.

A pre-Romantic, the idea that we are only ourselves when alone would have struck him as utterly bizarre. In 1794, when he wrote *Zoonomia, or the Laws of Organic Life,* it was plain as day that beavers only become clever in company:

> *'[Beavers are] possessed of amazing ingenuity [...] only where they exist in large numbers [...] while in their solitary state they shew no uncommon ingenuity.'*

The clubbable Enlightenment intellectual was able to see what Romanticism was about to occlude for the rest of us. (Coleridge hated the orangutan hypothesis, and I doubt he'd have cared for the aperçu about gregarious beavers either.)

Beavers are not the only species that only become clever in company. *The Entirely Accurate Encyclopaedia of Evolution* has shown the same thing happening among ants, bacteria (*Vibrio fischerii*), slime mould (*Dictyostelium discoideum*), nematodes, baboons, buffaloes, hominids and humans. In all these organisms, why is coming together conducive of brilliance? (Literally so in the case of the bioluminescent vibrio!) To show the same thing happening is not, alas, to have the foggiest idea as to how these species only get smart in a group. We saw this phenomenon in the very first section of the book (ants) so is there anything we've learnt from the rest of this encyclopaedia to help us to understand this phenomenon? Let's quickly review what we have learnt:

- Stonehenge General Hospital was missing its waiting time targets.
- We may be descended from Victorians.
- Unless you are lonely the birds will eat you.

Hmmm.

Anything else?

A recent paper published in *Ethology* found that of two populations of blue wrasse cleaner fish near Lizard Island on the Great Barrier Reef, the teeming population was more intelligent than the sparse and scattered one.[61] In fact, the teeming population was so intelligent that they were even found to be co-ordinating hunting strategies with moray eels.

The paper's authors propose that among sparse populations longer time intervals between activities make learning more difficult. This offers a clue as to why beavers 'in their solitary state [...] shew no uncommon ingenuity', and also goes some way to explain the geometric growth in the intelligence of large beaver colonies. Intriguingly, the same study also noted that the large, smart population of cleaner fish was located in a more complex physical environment. Now this could just be because their intelligence made them better at coping with a complex environment, less likely to get lost, say, but it might also go the other way too. Does a complex physical environment elicit flexible responses? Does living among intricate, complicated coral switch on dormant capacities in the same way that being in a large population seems to do? And if so does the complex lodge that beavers build create a feedback loop, whereby developmental plasticity and niche construction spiral upwards?

61 Sharon Wismer et al., *Variation in Cleaner Wrasse Coperation and Cognition: Influence of the Developmental Environment?*, Ethology (2014).

Maybe plasticity offers a clue towards solving Erasmus Darwin's beaver conundrum. Maybe if I hadn't wasted so much of the plasticity entry on an unseemly public spat with the other Robert A Newman, I might have seized hold of a golden thread to explain how nature selects for increased intelligence in a large colony of beavers when it's absent in a small group. Let's go back to plasticity, then.

When a few lacklustre beavers join a large colony, that colony can be said to be their new environment, in the sense that each beaver is environment to the other. This being so, their new beaver smarts can been seen as a novel plastic trait elicited by this new environment, no different from the way that daphnia put out a tail spike in response to the presence of notonecta in the pond.

...Or is it all much simpler? Perhaps the pure excitement of being in a large group and building a beaver dam together gives the beavers a contact high, and stimulates them into new ways of being. Perhaps it's like when people go on a mass direct action, or to a big gathering, and sometimes discover aspects of themselves they never knew existed, which turn out to be extremely influential over the future course of their life.

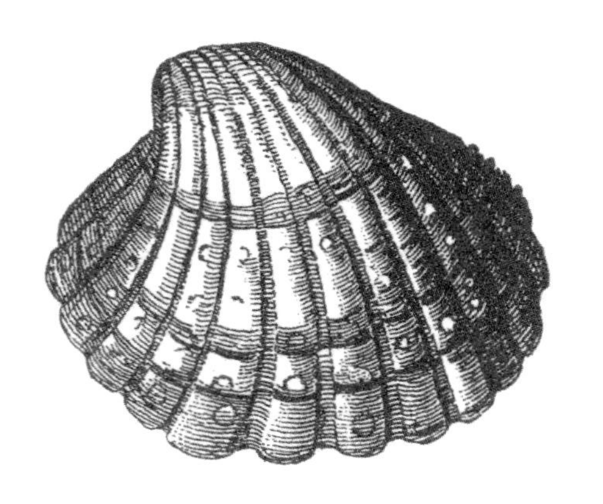

OUTRO

About a month after Marie gave me the Auden anthology I was on a train platform on my way to do a gig at the Common Place in Leeds. Re-reading the misfits poem, I got on the wrong train, which became a bus, which became another train, which 'experienced difficulties' and was shunted into the siding of an obscure branch-line station where it shuddered to a halt.

I ran up the platform, found the station guard, and told her that I was late for a gig. It so happened that one of the very few unselfish things I ever did in my whole life was a benefit for where she used to work. So she said,

'Wait back on the train. I'll see what I can do.' Then she literally ran off to try and sort things out for me.

Back on the train I realise that I am in the seventh circle of hell. Not only am I late for a gig – one of the most sickening, shame-filled feelings I've ever known – but the men in the compartment have struck up a conversation and discovered to their mutual delight that they are all management consultants. They're all bragging about who has been the most courageously unflinching in the course of efficiency. One of them's talking about how, when this train gets moving again, he's going to be recommending that this particular firm make 15% of its workforce redundant. Another feller's talking about how he's going to be recommending a firm sack the entire workforce and then reemploy them on short-time contracts, someone else is talking about how he's going to be recommending mothballing the whole company and shipping it overseas. I said,

'Excuse me, did any of you gentlemen trouble yourselves to inform the respective workforces of what sweet plans you have in store for them?' One of them looked me in the eye and actually said,

'Oh no, no, no. Bad for morale.'

Bad for morale! Here I am in the epicentre of this dog-eat-dog universe. Why fight it any more? After all, these are the people controlling everybody's lives.

Just then there's a knock at the window. Conversation drops. We all look up. But it is me the station guard is beckoning with her coral-red fingernail, me to whom her coral-red lipsticked mouth is miming the words: Come on. I grab my gear and join her out on the platform where she says:

'Okay, there's another train line 25 miles from here on which trains are still running, and there's a train coming in one hour that will take you where you need to go.'

'That's no good to me!' I shrieked. 'I can't afford a 25-mile cab fare! And what if all the time we're talking here this train suddenly pulls away? Oh, what am I gonna do?'

'I've spoken to the engineer,' she replied in a calming-down-a-dangerous-nutter tone of voice. 'This train isn't going anywhere for another 48 hours. I'm just going to announce that fact on the tannoy right now.'

'No, no, don't do that,' I said. 'Bad for morale.'

'You think so?'

'Could lead to sabotage.'

'Okay then,' she said. 'Come with me.'

She led me across the goods yard, we went round the back of a single-storey red-brick building, and there, beside a low platform, sitting on the track, was a see-saw hand-operated railway trolley as used by Laurel & Hardy, Charlie Chaplin and Buster Keaton.

'Just keep pumping till you get there,' she told me. 'You won't get lost!'

I cannot describe to you the feeling when I first plunged down on the T-bar. The sense of transgression could not have been greater if I was plunging the detonator to dynamite a vacant office block. And sure enough, as the wheels creaked and turned and I began to trundle down the track, the office blocks fell away, as did the car parks and shopping malls, and I found myself scudding through fields of brassica under wide-open skies. But I kept stopping to pick up passengers. To some kids who'd been crabbing in a nearby stream, I said,

'Get on board!'

To some Eastern European fruit pickers just knocking off at the end of their shift, I said,

'Get on board!'

The fruit pickers climbed aboard and spread out a gingham cloth, and set out bread, olives and wine. Soon I was sharing the pumping with a Bulgarian woman.

'This scenery is very beautiful,' she said, less puffed out than me. I followed her gaze and seemed to see, as if for the first time, the purple loosestrife, rosebay willowherb and the tall hornbeams.

'Yes, I suppose you're right,' I replied. 'It is rather.'

'My last boyfriend,' she said. 'He did not love countryside.'

'Oo-arrr!' I wurzled.

It was getting dark when the train station came into view. I was going to make the gig on time. As we pumped that railway trolley the last stretch of track, unusual moonlight revealed the earth to be some other more favourable planet than the place I thought I knew, where many more things are doable than ever we dare to dream, but where we have spent so long believing that dog-eat-dog economics are a law of nature, and ourselves lumbering robots controlled mind and body by selfish genes that when we finally shake off these popular superstitions we discover that time is running out. As our rackety raft of misfits rattled down the tracks, it suddenly seemed actually possible that mass action could halt the slide into irretrievable climate breakdown, but I also knew that, in the words of the great philosopher Bertrand Russell:

'We had better put a wiggle on.'

THANKS & ACKNOWLEDGEMENTS

Helen Sedgwick, Gill Tasker, Leonie Gombrich, Rhodri Hayward, Jonathan Harvey, Mary Midgley, Clare Alexander, Ed Smith, George Monbiot, Katharine Ainger and Vesselina Newman.